Pyramid in Space

Bill Wheatley

Pyramid in Space

a twentieth century life
viewed from a distance and rendered
as a collage of scenes

"Pyramid in Space," by Bill Wheatley. ISBN 978-1-951985-81-3.

Published 2021 by Virtualbookworm.com Publishing, P.O. Box 9949, College Station, TX 77842, US.

Pyramid in Space is a collage of scenes based on actual events, so it can be called “true-ish,” as Dr. Gillies used to say. However, I have changed the names of most of the people because the past, even if recorded faithfully, is its own foreign country and never feels like the days we are presently living through. So the characters described here are both themselves as they once were and people who may have never existed.

ONE

Seventy is when I started feeling old, started feeling as if I'd passed over some invisible line and this is what it was like to be old. I still work hard in the garden, even prying those big rocks out in the spring after the fiasco of Ronnie digging a hole with his backhoe and burying most of them. Driving back and forth over the garden he pounded down the soil so hard we had to hire Charlie to plow it, then till it again, and the plowing brought up a new batch of sizable rocks that had been hiding under the topsoil. I'd sworn off levering them out, thinking if I kept hurling myself into physical tasks I was going to get hurt. Then I'd really regret it. These are all things I wouldn't know if we hadn't started gardening seriously twenty-five years ago.

Becoming a gardener is just about the last thing I thought I'd be doing, and enjoying it too. That discovery may not be a bad starting point. Begin with the unexpected, modest as it may be, as a reminder that almost nothing turns out the way you thought it would. After a certain age no one is living the life they expected to live. So that most of us have a constant clanging in our ears from the dissonance between who we thought we'd be and who we are. This experience can only be made more common by growing up and growing old in a technological culture. It makes us all a little dizzy. There are exceptions, the few whose lives turn out to be just what they expected. As they grow toward completion, they develop a smoother irony, as if catching a glimpse of themselves in the mirror they think, "Yup, just who I thought I'd see." That doesn't mean it's enough; it might mean your life lacked surprises. However it turns out, something in the organism always wants more. The religions of the east seem to be honed to master this spiritual greed, to teach sufficiency and repletion instead of excess. It's not our American Way, here where the pick-up trucks get bigger every year against all logic or reason just to flatter our vanity, our basal cell desire for power. Or for its illusion.

From the top of a pyramid, especially if it's out in space, the unlimited view is exhilarating, the opposite of being stuck in one place on the ground, living one short life, teased by all the possibilities that won't happen, by all the experiences you won't have, the people you won't meet, the thoughts your brain has too few cells to comprehend. The last one bothers me most. There are things I might understand if my brain were bigger, with more neurons, more interconnections. Not only that, my brain is shrinking. "I'm sorry--it happens to everyone," the attractive female neurologist said, apologizing for bringing bad news. So not only are there things I'll never understand, an earlier version of myself had more mental muscle than I do now. There are many doors that will stay shut for the duration of my short forever, hiding mysteries I can never know. It gives me a pang to be reminded there is so much I'll never know, that I'll never know how this all turns out either. I know that already; it doesn't "turn out" like a neatly told story.

There is no shapely narrative. It just rolls on with its blend of the comic, the tragic, and the absurd until it fades out or blows up or is transformed into something no one could have anticipated.

2. Escape

It isn't easy. I couldn't just wish my way out here on this pyramid. I had to live seventy-plus years, go through the usual trash, enjoy some distracting rewards, yes, that too, and how, exactly, I got out here is not something I precisely understand. Perhaps I will, if I get some more time and plug on. My heart's been fixed, I've been saved by science (at least for now), so maybe I will get a few minutes more.

The view I really want to see, to have revealed to me, is--

Maybe I shouldn't say it out loud. Or write it out in black and white. Certain things go away when you name them. They want to be unnameable. As Beckett named his book. Or as the ancient Hebrews called their God by his initials, YHWH, too awed to speak his name. We might represent the mysteries of where our universe came from by the initials CMBR for the cosmic microwave background radiation still glowing out there on the furthest horizon after the big bang, almost fourteen billion years ago. A way of reminding us that if we really knew what we were looking at we'd be too awed to speak. Which wouldn't get us anywhere. Better to be impressed by ourselves and able to speak. If we want any progress, that is. (If that's what it is.)

I would like to understand. I can say that. We'd all like to understand, I think: we'd like to know. It isn't going to happen. I've learned enough to know that for sure. The answer is scientific and something I can see quite clearly from up here. I mean I can see that I can't see, won't see. Even from up here. Out here is what I should say. I think I'm more out than up. Either is just a convention and one I couldn't tell from upside-down anyway. Which is an interesting idea I must look into when I get an idle moment.... So, life forms (we don't know how) or emerges and off it goes wriggling or crawling or oozing according to some force we don't understand, although we give names to it that reveal assumptions (life force), yet insist whatever it is it's directionless (this is gospel to the professors of evolution). In other words we can say what it's not but not what it is. This for me personally is a problem. I know why scientists insist that the development of life, evolution, has no direction, no intention, and I have to agree, yet I can't see why they're so sure they can prove this negative. They need to assume it because the moment you say evolution has a purpose or direction you tumble into the waiting arms of religion, or something very much like it. The Teleological Fallacy. Believing all this is heading somewhere. Yet proving a negative is notoriously hard, impossible--how can you prove every case wrong?--and it's a tenet of science that an argument shouldn't be based on this kind of flawed logic. But I understand. It's taken so long and been such a fraught bloody battle to liberate science from superstition and religion it's easy to see why science insists on saying the development of the natural world is merely ongoing, but not going anywhere in particular. The apparently increasing complexity of, say, primates like us and our ancestors doesn't mean--what doesn't it mean? It doesn't mean we know what nature is up to. It isn't up to anything. The progress we think we see replicated in our own innovations is just that, our innovation. Which in this deepest sense cannot embody any intent but our own. Human progress is our own invention and we can be proud of it but we can't say we've invented any new physical laws. We're still creatures playing someone else's game. And so much the better. I mean, imagine if we had to make it all up ourselves.

3. Intuition

The counter-intuitiveness of so much in nature is upsetting. I want the reality to confirm what I see. Why does it matter? The question seems too simple to ask: so what if things aren't what they seem? So what if they're something else. The earth isn't flat, it just looks that way. When you get far enough away, like viewing it from a pyramid in space, you can see the earth is round, spherical, a globe. It's that big blue dot right out there. There are plenty of these earth-looks-flat phenomena all around to make you think, now and then, you've gotten something wrong. Maybe you've gotten everything wrong. What I also have noticed is that once you get the revised picture, you think, okay, now I can trust my eyes. Your credulousness is continually renewed. I know the earth is round even though it looks flat, and I know pyramids don't move. Yet they do, not all the way out to space, maybe, but they are still on the move, "both with common motion and with their own," as a French observer once said.

Is there any point to living in a state of doubt so complete you never trust your eyes? Of course not. So this is where the trick comes in. You've got to pick your spots, choose where to draw the line between what you think you know and what you know you don't. That means checking things out, sometimes in detail. In our own time, you might learn to your surprise, as I have, that there is no general agreement among physicists about the nature of reality. Quantum theory, which supposedly describes the behavior of certain subatomic particles, can't say with confidence how they really do behave. One can only predict probabilities, not certainties, and even about them there are mysteries. Once again, comes the question--does it matter? Someday it will, says physicist Adam Becker. It doesn't matter if the earth is flat if you never move faster than a walk or travel further than to the next town on your donkey. But if you're launching a rocket into space you want to know the earth is round or rounded or elliptical so you can tell where the rocket is going to come down. Or if it stays up there, what orbit it's in. So, yes, how subatomic particles behave is important to know, or will be, even if no human being has ever seen one.

We take it on faith that they exist. Or rather that the story physicists have made up to describe subatomic actions are plausibly "true" representations of whatever it is that actually happens. These descriptions, metaphorical as they must necessarily be, are what we believe in. Cold comfort, to be sure, compared to the warm assurances our ancestors lived with. "It has taken eight centuries to replace the conception of existence as a divinely composed and purposeful drama with the conception of existence as a blindly running flux of disintegrating energy," said Carl

Becker (unrelated to Adam). Nevertheless, what we're left with, science instead of faith, doubting more than believing, is not nothing. We have the satisfaction of knowing how things, some things, really are. (Well, we think we know.) And science, in its medical form, can save your life. It has saved mine. Which is more than any mythical supernatural being could do. What we don't have is the very different satisfaction of believing one of those mythical supernatural beings cares uniquely about us. We have lost the experience of sharing with a large group of people the feeling they believe in the same exalted spirit that we do. Although maybe you get that experience at a concert, or watching a performance, or sensing some other work of art, or even a sports event. It doesn't connect you with the immortal in the traditionally promised way, but the sensation can be spiritually elevating, it can transport you out of your petty limited self like nothing else.

4. Prisoner of gardens

Gardening connects you to something greater than yourself too but in a very different way. Soil, weather, pests, disease, weeds, and vegetables. The garden changes over time but holds us tight. As Kate and I get older we wonder how long we can continue to cultivate such a large one. It's only ten thousand square feet or so, much too small for the ambitious name farm. Judged by that, we're dilettantes. Nevertheless, we do everything by hand except for Charlie's fall tilling with his small tractor. Besides age, climate change is beginning to press against us too. A few years ago the prediction for the effects of global warming in the northeast was "warmer and wetter," as our neighbor Emery said. He works at Harvard Forest where tracking the weather is serious. It's an example of how unpredictable climate change can be, that instead of warmer and wetter, we're also getting colder and wetter. Winter is lasting longer, sometimes. The warming of the Arctic Ocean has melted so much ice that the jet stream is looping south, losing its grip on the cold arctic air and allowing it to pour down on us from Canada, all the way to central Massachusetts. This year we've had four nor'easters in March, and two snow "events" in April. Not normal. One year is not a trend, blah blah blah, but this is a new phenomenon and it can be explained by the melting Arctic ice cap which no longer exists. There was a day in February when it was 15F here in the middle of Massachusetts and 35F at the North Pole (anyone for a swim?).

5. Bud's career

"Anyone for a swim?" is just the kind of slangy, irreverent response Kate's father would make to a globe-sized problem threatening his very own civilization. Bud came from a well-to-do family in a small Missouri town, studied art in college, and got a job as an illustrator at an advertising agency in Fort Madison, Iowa. After Pearl Harbor, when the war started pulling everyone into its orbit, Bud enlisted in the Navy which sent him to Officer Candidate School. He fathered two children while on furloughs from serving on a minesweeper in the Mediterranean, and when he came home instead of returning to his advertising job, he started working for his father-in-law. Eventually he took over the small construction company that mostly built roads.

He was happily suited to life as a human being. His wife often said, "Bud would have been happy doing anything."

Bud worked in the warm months, overseeing his small crew and often handling the equipment himself, and when it was too cold to lay asphalt he worked on his hobbies. He fished and hunted waterfowl, enjoyed cooking and the camaraderie of his friends, and the parties he and his wife went to in their small town weren't over until he had led the singing of "McNamara's Band" to send everyone home. He seemed to fulfill Robert Burns' prescription for the ideal human person: "The social, friendly, honest man, what e'er he be/'tis he fulfills great nature's plan, and none but him."

Sitting across from Bud, both of us with a drink in hand, he in his stuffed leather chair with his artworks portraying ducks and geese hanging on the paneled walls, a shotgun cabinet beside the television, his golden retriever sitting next to him with a canine smile on his face, I often had the same thought: Bud could never die, he's too well-suited to life. He is so positively anchored here in the middle of the midwest, that the ordinary forces of decay and disease seem irrelevant. His good health and baseline happiness is a kind of feedback loop that keeps itself going, as if that's what it's meant to do. I had this thought even though he had a sizeable overhanging belly and had only quit smoking cigarettes recently. Those seemed to be merely features of his perpetual motion. Because he took life as it came and wasn't preoccupied with his health, he never reminded you of mortality. He was as solidly rooted to his place in this world as a pyramid. Then with no warning, there was a rip in this seamless fabric; he got cancer and died a few fleeting weeks into treatment. It was a lesson in how swiftly a single new fact can change everything. That solid presence, even at eighty, as if he belonged here so thoroughly he could never be subtracted, could vanish with the same naturalness that had sustained him before.

The jarring juxtaposed images of Bud alive and Bud dying come to me now provoked by the evidence that we're moving helplessly closer to an ecological cataclysm, possibly already past the tipping point as Greenland and Antarctica melt. All of us, not just Americans who've heard the warnings for years but everyone everywhere who will be affected whether they've heard of it or not, to say nothing of the rock-headed deniers, all of us are Bud sitting in his easy chair in front of the TV, drink in hand, golden retriever as comforting as a therapy dog, watching the televised world go by without seeing anything that might disturb our equanimity. With so much inertia, how could this ever change?

Since the trait is universal it seems as if we must have been "designed" by the improvised motions of evolution to be complacent. Becoming too easily alarmed can be a waste of energy and a distraction from tending to the necessities of living. The person who is always warning the sky is falling is regarded skeptically by every culture. But why does this trait so often go too far in the other direction, paralyzed by complacency? Is it because the people most heavily invested in things as they are have the most power and assets? They're the ones with the most to lose in case of calamity too, of course, so it looks as if there's no simple, clear answer.

Whatever the reason, this is what we're left with. It's the summer of 2018 as I write, California is on fire, our garden is flooded with rain, the North Pole has had above freezing temperatures in four of the past five years, and we continue to pour carbon into the atmosphere warming the planet even more. We have a cataclysmically changing climate and a dominant species ill-suited to taking action that would protect our future welfare. The dark view says these are signs our civilization is spiraling down toward collapse. The more optimistic view--well, that was represented by Bud.

6. The center of the universe

On the afternoon of the seventh game of the 1960 World Series they were lined up three-deep at the soda fountain in John Betts Pharmacy, sitting and standing and leaning around each other as if composed for a Norman Rockwell painting. The kid behind the fountain--the soda jerk--could see nothing but a wall of heads staring past him at the small black and white television set placed temporarily on a cleared space above the ice cream freezer. Standing in back of the crowd, taller and heftier than anyone else, wearing his standard issue white shirt, dark tie, and black suit trousers, was Mr. Betts, his bullet-hole eyes also aimed at the television screen. When you spoke to John Betts or got his attention, those eyes stabbed you to hold you in place, heard your request then answered in his rushed breathless growl. With his size and carriage, his slicked back hair and oblong face hung with jowls, John Betts resembled bandleader Paul Whiteman and enjoyed conducting business in his drugstore with a certain similar flair. When the transaction was done he gave you, if you were a man, a full-throated comment or pleasantry or joke to leave on, his face often crumpling with merry laughter as his cheeks glowed red. With women he was protective and patriarchal, his voice so confidential he had to lean down to be heard, his manner assuring them he was discreet, no, secretive, filling any prescription for their health or personal needs. At this moment, however, the pharmacist's attention, and his immediate fate was held in a place he dearly resented--someone else's hands. Or to be more precise, in the bat in someone else's hands, the bat he might or might not swing at the next pitch thrown to him. John Betts, as I said, despised yielding any control over his life to anyone except his wife to whom he had ceded authority over areas that didn't interest or were beneath him. He also despised the IRS for its intrusions into his life, teenage delinquents for malingering on the sidewalk in front of his pharmacy, and anyone who altered or obstructed John Betts' way through his daily routine. Yet despite this bone-deep abhorrence of dependence, every day John Betts placed wagers (never using the offensive and unfortunate homonym with his surname) on the outcomes of sporting events over which he also had no control whatsoever. So he lived every day with his self-satisfaction and pleasure constrained by the imminent news of outcomes he couldn't control at Vernon Downs harness racing track, or the games in Major League Baseball, or the basketball and football games in the National Collegiate Athletic Association, or the National Football League, or the National Basketball Association, or fights sponsored by the World Boxing Federation. He also played card games of poker and blackjack down at the Elks' Club Lodge, where players backed their luck and competence with cash. A game of cards or a sporting event without money wagered on it was pure voyeurism, an idle amusement, a kids' game; a real man entered the fray with his own worth at stake, and shed tears and sweat blood with his competitors.

The bets John Betts placed on these events were not legal so had to be facilitated by an independent, unlicensed, itinerant banker named Vito who stopped by every afternoon to pick up an envelope which he would "take to the track." If anyone hit a winner, or the Daily Double, including the women who worked days at the soda fountain and often split a two dollar bet, Vito was there the next day with another envelope to pay off. Ultimately, some of the wagered money probably found its way from Vito's discreet hand into the maw of organized crime, an unsavory thought tempered by the fact that Vito himself was the smallest of small businessmen, a type extolled by John Betts as the foundation of the free enterprise system. To provide them employment, and the occasional appreciative tip, could not be considered a crime except by prudes, the kind of thing the State should wink at, concentrating its energy on threats to the safety of law-abiding tax-payers from violent felons and the hoods who hung around street corners even in respectable neighborhoods like Eastwood, scaring women and the elderly from crossing the sidewalk to enter a five-and-dime or a pharmacy.

John Betts was a Republican, he was the Past Exalted Ruler of the Elks Club Lodge of Syracuse, New York, a graduate of the Albany College of

Pharmacy who was licensed to dispense prescription medication by the State of New York. Besides being a self-employed entrepreneur and small businessman, he was married and childless, a drinker of J&B whisky and a smoker of Bering Double Corona cigars. He frequently picked up the check for the table, never counted his change, and always left a generous tip, unless he'd been treated with indifference or carelessness, in which case he might even speak to the manager in his deep, confidential voice, waggle a long fat finger, and dispense a brief lecture on a business's obligation to take care of the customer. Mr. Betts strode to work every day through the greatest neighborhood in the greatest city in the greatest country in the world, wearing his black Homburg and black wool topcoat (in season) and waving a brisk hearty welcome to all.

In a further paradox, John Betts, lifelong resident of the State of New York, admirer of excellence and skill in any arena, and covetous of the power gathered to himself, hated the Yankees, the most successful sports team in America. The prejudice went back to his youth. The New York Yankees had had been the nearest professional team and all the boys he knew were fans. Because of his native individualism, his desire to be original in his realm, and to control his own tastes and enthusiasms, he couldn't root for the same team as everyone else and remain true to himself. He decided to pledge his allegiance to the Brooklyn Dodgers. John Betts knew that the gambler who wins is as cold blooded as Josef Stalin. As a player of games of chance he regarded himself by nature and skill as a winner--a magnanimous one, who could win the pot cleaning you out then hand you a sawbuck to get home, shaking his moneybag jowls in rejecting your promise to repay; and if you did bring it up again, John Betts would have forgotten it all, truly forgotten it, or he would refuse to remember, offended someone thought his generosity had to be repaid. It didn't because he was meant to win, it was his right, his destiny. It couldn't be his bad judgment or bias that put him on the wrong end of a wager but a perversion of luck or a fixed race or an indifferent player on that given day. He was also susceptible, as a human being, to the insidious feeling that one wager, his own, either made or not made, could in some undefinable way, affect the outcome of an event. (It had happened to him!) He was also an honest man. Scrupulously so with friends, associates, customers, but not so punctilious with the government, whose own excesses and corruption were contemptible. He also believed he was honest with himself, as honest as he could be without losing his integrity. "Those damn Yankees have been killing me all year, Eli," he had said before the Series began, making a rare confession to the young kid behind the soda fountain. "I hate 'em, but I've had such a bad year I've got to take 'em in the Series and get even. Gotta do it."

Exhaling the phrase in a sigh, Mr. Betts settled his heavy torso on a stool and was stirring the coffee he always paid for at the soda fountain in his own drugstore. His lower end was afflicted with a painful fistula, which he favored by putting all his weight on one cheek and supporting the cantilevered buttock by bracing the extended foot against the base of the soda fountain.

I had only worked at the drugstore for a month but had quickly learned that any comment uttered by Mr. Betts other than a direct question ("You ready to close up now, Eli?") not only required no answer but if you did answer you'd be ignored. "I know I'll regret this," John Betts repeated a few days before the World Series began, showing how much it was on his mind, "but I haven't any choice. They've been beating me too badly this year. I've lost a lot, Eli. A lot!"

This kind of confession was entirely new to me. I didn't know that even a man as self-assured as John Betts needed to express his doubts aloud now and then. I also didn't know that revealing them to someone as insignificant as the teenager serving him coffee from behind the soda fountain in his own store meant his misgiving had hardly been spoken at all. Somehow I did understand that Mr. Betts' using my name to address me was not to encourage familiarity but was, in its way, affectionate (Mr. Betts was proud of the "good" kind of kids he always hired) and because it was affectionate it also pledged me to secrecy. It was the other half of a pact, and naming the partner swore him not to reveal John Betts' decision to recoup his money by siding against himself. To bet on the Yankees was to descend from the high ground, to forfeit some of his individuality, his independence, his distinction. It was the opposite of gambling, the very thing that his nevertheless Republican, Christian, American self wholly disdained: the safe, the easy, the trite, the trodden, the way of the herd, the place every other poor sonofabitch could go.

So now John Betts stood in the back of the crowd filling his drugstore, passionately waiting and hoping the batter at the plate would make an out and redeem his gambling losses for the year, and also feeling that his position was false and he should be rooting for the Pittsburgh Pirate at the plate to get a base hit, or even an improbable, impossible, homer.

Bill Mazeroski was a gifted fielder who was belatedly voted into the Hall of Fame by the Veterans Committee in honor of all his baseball talents except hitting. He was leading off the bottom of the ninth with the game tied at nine runs apiece. Just for the Pirates to have stretched the Series to a final seventh game had been against the odds because the Yanks had been hammering Pirates' pitchers all during the Series. When the Yankees came back to tie the last game in the top of the ninth it felt inevitable they'd win just like they always did. It was happening again. Their victory was somehow intended. John Betts had sweated out the Series but now, finally, it looked as if he'd made the right choice. A gambler entering the contest at this point would have laid odds against the Pirates' winning. With Maz standing at the plate, waiting for the first pitch, John Betts felt a flashback of the torture he'd been through the whole perverse season. This team did not belong here. Three more outs, then into extra innings where the Yanks could finally put away these resilient but inferior Pirates.

The first pitch was a ball. On the mound, Ralph Terry got the sign from Yogi behind the plate, set himself and threw the next pitch. Mazeroski swung. He hit a long, deep drive to left field and the faces along the soda fountain froze into Norman Rockwell poses of astonishment and surprise when they realized that what they were watching on that little television screen was a dramatic reversal of all expectations, that the mighty could fall, David could rise up and slay Goliath again because the gods of baseball, like all deities, moved in their own mysterious ways. The struck ball had sailed out of the stadium and Mazeroski was running the bases, taking what seemed like forever. He was mobbed at home plate by his teammates. John Betts threw his arms in the air and shouted "God-*damn!*" loudly enough to be heard over the groans and cheers filling his store. "Shut it off, Eli!" he commanded, pointing a long, fat, quivering finger at the condemned television. Then he retreated quickly behind the glass-topped wall of the apothecary.

Although the loss of money hurt badly--more than twenty percent of his income that year--there was something else that hurt far more. This was the year, if he had placed his money on the team his heart and soul wanted to win, that John Betts could have enjoyed both a monetary and a fan's reward. But he had broken his personal code, stepped out of his own firm character, and he had done it venally, merely to make back his money. You didn't gamble just to win a few bucks. You did it to show you were able to discern the motions of invisible forces before they unfolded; even, in this way, to have some degree of control over them by sensing their direction before they moved. You proved your faith in your own powers as a member of this priesthood of seers by having the brass balls to back up your predictions with cash. So when John Betts lost by playing the arch enemy to win, the moral was so obvious it was embarrassing. It was humiliating, and it was shameful.

Plunged deeply back into character by the event, the bullet-eyed victim of his own choice, he took his loss like a Past Exalted Ruler, and when he recovered some months later even began telling the story on himself in his hushed confidential voice, doffing his hat to luck, to fate, and putting its relationship to him smack in the center of the universe: "What do you bet if I'd taken the Pirates, Maz would have struck out!"

7. One trillion

A physicist recently said he thought there weren't 100 or even 200 billion galaxies in the universe but one trillion. And how many stars are there in a single galaxy? A hundred million? Multiply that by one trillion. All these numbers are so far beyond a human brain's capacity to comprehend that we might as well be plankton or krill instead of "intelligent" primates contemplating them. Perched on the peak of a pyramid in space for a better view doesn't really help very much if you're trying to understand these extreme dimensions. At least it didn't when I tried it. All I could see from out there were more stars. If I turned toward the earth it looked huge, but the next time I looked it wasn't much bigger than the moon looks from earth which told me the pyramid I was standing on must be moving--and moving in some wobbly, non-orbital path. Or maybe it was just another special effect of the imagination. Before I got turned around again, lost in space a new way, I had a flash of memory that wherever one is in the universe--on top of this particular pyramid, for instance--is the center. Does that make sense? (None of this really makes sense.) Wherever you are is the center. Because there is no center. So you can't ask how far is it to the edge of the cosmic microwave background radiation. Yes you can because that's how they calculate how long ago the Big Bang happened. Yet maybe the CMBR moves when we move, so it is always as far away from everything else as anything can be. Is that the answer? The edge of space (there is no edge; if it has a shape it's curved) is always expanding. Accelerating away from us. Until it stops, reverses, and begins contracting toward another singularity of nearly infinitely tiny mass like the one preceding the last Big Bang. If that's what happened. It's just a guess. It's what they say happened. None of cosmic history is written down. There's no Bible or Bhagavadh Gita, no origin story except the sketchy one cosmologists have doped out from staring at the data retrieved by their telescopes. Ripples in the fabric of space are the evidence, the "sound" of a cataclysm in the early universe. Everyone seems to agree there was nothing and then there was something. But what was the nothing that something emerged from? That cannot be the whole story. Or maybe you'd say that's the beginning of the end of the story preceding the one we're living in. So to speak. It's beyond the reach of our understanding. Mine, anyway. Although mine is flawed in a peculiar way. I try too hard to understand, to make it real.

Nowadays the answer seems to be there's more than one universe. In fact, they say, there is an endless supply in a "multi-verse." This has the with-one-leap-he-was-free feeling to me, but then I'm not a physicist. The multiverse is usually described as infinite but I wince when I hear the word infinite. To me it has about the same meaning as "God." That may be because I was raised in the church and every week inhaled a dose of Presbyterian metaphysics. But think about it. No human mind can conceive or contain the idea of infinity. It's an idea too big to fit into our skull. So I don't think we should talk about it as if we're familiar with it. As if it's something we understand. Better to say there is something beyond our understanding but we don't know what it is. Or there is nothing beyond our understanding but we don't know what it is. Or isn't.

8. Dimensions

One of the things Montaigne does in his *Essays* that makes him unusual is to describe himself physically. This is more rare than you'd think. Writers describe their characters' appearance, often supplying a trait or two that is meant to be a clue to their personality in the case of fiction, or simply a handle for the reader to hold on to, but frequently even in non-fiction the simple facts about a person's physical being are, surprisingly, not given. You could argue that since fictional characters are imagined, the "simple facts" about their physical being don't apply because they don't really exist (which might lead to a conversation about the dangers of making fiction too real, a speculation one might really need to speak French to express). But I think physical form is so important to all of us that whether a character is real or imagined a writer should have a good reason for not supplying the stats about how much space they take up on earth. A person's height and weight are basic to their identity. Anyone can tell you those numbers about themselves, just like their age. When Mom was ninety, I took her to the doctor where the nurse measured her height. "Five-three," the nurse said. "Oh, I'm taller than that," Mom said, as if she'd been shortchanged. She'd always been a little proud of being five-seven. "Not now you aren't, honey," the nurse smiled. She had probably heard that a lot from old folks.

Montaigne tells the reader he is below average height, thickly built and not good at sports. No spin, no archness, no defensive or assertive tone to the voice. Just facts. He doesn't use the info as fodder for psychoanalysis. Well, this was written about 1580, so the sensibility was different from ours. Besides, unless Montaigne is dissembling, his purpose is not to compose a self-portrait for anonymous readers; instead, he tells us, he is writing for himself and for his family and friends. Then why does he need to disclose his size and lack of athletic ability--don't they already know? That's a good question. It could be merely a starting point, a way for him to orient himself toward creating this self-portrait, although it comes a good half way through his essays, so it's not his literal starting point. Or it could be that he is secretly hoping for more readers than his family circle and he reveals this wish without quite noticing what he's doing. Since it's more accurate to say a writer often doesn't know where the urge to write any particular piece comes from (unless he or she is paid for it), what the writer says is the reason--if he or she isn't too coy to say it out loud--may itself be a fiction of some kind even if it's intended to be a factual report. And of course he could have been dissembling which is a common trait of people who sit in a room alone and try to describe themselves. (A review of a recent biography of Montaigne said his retiring from the world to contemplate *la vie humaine* was a pose and he was writing to advance his political career as much as to examine the human experience.)

Whatever the case, it's interesting that Montaigne presents himself as a physical being and does it with apparent candor, when even in our confessional age writers can be reluctant to stand before a reader in physical form, as if revealing that limited mortal being might somehow fatally undermine the desire to be known by giving away something the writer needs and the reader can do without.

For the record, since I brought this subject up (Montaigne brought it up first), I'm a size medium of average height, slenderly built, a middling athlete, not gifted but not a klutz, with a certain degree of physical grace, at least before I stumbled into my eighth decade and lost what had been a pretty good sense of balance. Having said that, I suddenly feel--not naked, but in some unsettling way exposed, or reduced in size. Metaphorical size, I mean. Am I no more than these modest dimensions? That may be the source of the reluctance to give one's vital statistics, something like superstition. (We all have our secrets, as Maddy said.) A poet wouldn't tell you these prosy things. (They might; poetry can be anything nowadays.) It also feels like a death-stroke to the imagination. The plain description is so reductive, it strips my personality, erasing any of those qualities that make me individual instead of statistical. If there is anything special about me, it's gone now. Gone so abruptly I find myself on the other side of the argument. If that's the effect of vital statistics on a character then a writer shouldn't divulge them. By being too specific, they hobble--even undermine--the effort to portray a real human being. But I don't care about being a real person here. This is a picture of something else. What that something else is, I can only define by writing it. Anyway, I'm not going to go back to delete my own data. What's done is done. That was the rule when I wrote Joe's book and why should this be any different? (It's a different kind of book.) As Joe said, giving me permission to use anything he said: "Fuck it, I said it--it happened."

9. Tomatoes

Some of the most decisive events seem the least intended. When Kate and I lived in Cambridge, the man across the street saw her in the tiny fenced yard by the side of our shabby, rented duplex apartment one late spring day and came over and asked if she wanted some tomato seedlings. He had a few left over after filling the yard on the other side of his house with new ones he'd started indoors. Kate said no. There wasn't enough sun and she didn't want to have to water them. "I'll do it," Gabe said. "There's a good half day of sun, it's probably enough." He was likely feeling a little sentimental about the seedlings. You do, I know now--after you've started growing something from seed it's hard to throw it out. It turned out Kate was right. We didn't get enough sun, especially during what turned out to be a rainy August, and Gabe's wee plants never produced fully ripened fruit. "Next year we'll start 'em sooner," Gabe said, in the try-again spirit automatic to veteran gardeners. But the next year we were living in the country, another unexpected change, and a few years after that we were raising more than eight hundred tomato plants from seed every spring and growing all our own vegetables. Tomatoes have always been our lodestar crop for eating, cooking, and preserving, and when we read an article about heirloom tomatoes in the New York Times we were seduced. Just the names of these old varieties made you want to grow them: Brandywine, Black Prince, Aunt Lillian's Yellow, Orange Banana, Green Zebra, many of them too big or soft skinned to travel in trucks to market. So over the years they had been replaced by hybrids with tough skins and firm flesh instead of juicy pulp, raised for their virtues as market commodities, not for their taste.

What we couldn't eat or preserve ourselves, we began selling from a roadside stand and we made enough money to pay the garden expenses and buy heating oil for the winter. Without quite knowing what we were doing, we had backed into the role of serious, nearly full time gardening, for Kate anyway. A cultivated third of an acre at our max wasn't a farm, but it was more than a vegetable patch, and once begun we embraced it, almost helplessly it seemed. We discovered unsuspected things about ourselves and about plants, learning what it was like to live close to the soil, wholly dependent on the weather. Most any book or magazine article recounting the journey from the city back to the land sounds like what we experienced, except most people are more conscious of what they are doing than we were. They have a plan; we just had a garden which we kept enlarging, square foot by square foot, the servants of a nascent manifest destiny which nurtured an urge to keep growing more, more, more. Instead of a philosophy or plan of action, all we had were two stipulations. We'd grow organically (whatever that meant) and we wouldn't buy powered equipment, we'd only use hand tools. I didn't want to have to maintain or repair anything with an engine, like a rototiller. Instead we hired a neighbor with a small tractor to plow and till the ground to be cultivated. We also paid him to cut and bale our field of hay. That gave us material to use for mulching.

"Organic" is a much abused word which often doesn't define anything rigorously enough to be meaningful, especially now that so many commercial growers and chemical manufacturers are crowding under its halo. For us it meant only using pesticides approved by a local sustainable agriculture group. The permitted ones were less potent than the chemical killers, and we also used them sparingly. The "organic" fertilizer we used was also weaker than the turbocharged products, and we started using compost and cover crops to improve the soil. Every aspect of the enterprise was new to us, but we weren't kids and knew how to learn from books. Local growers were glad to give advice too. Mostly we learned you had to pay attention and do things according to the plants' needs, not your own. We rotated crops and drew maps to keep track of what had been planted where, gradually reduced the mistakes we made and increased our knowledge of what worked, sticking close to the advice in two gardening books we treated as our bibles. They rarely disagreed with each other--a good sign--and one was by Shepherd Ogden, a commercial grower in Vermont who'd first learned about gardening from his grandfather. He had weather like ours, an important factor, and plenty of rocks in his field, just like us too.

Our policy on using pesticides was also one we had backed into. When I found worms in the first carrots we grew, I drove down to the local Agway and bought a small bottle of Sevin. I didn't read the label until I got home and when I did it sounded so toxic and powerful ("add two drops to a gallon of water"), I didn't even open it. We weren't growing our own vegetables only to use the same pesticides the produce in the supermarket had been drenched with.

The experience with Sevin became a touchstone and it illustrates that as the garden grew the reasons we were investing so much time and effort in it became more apparent. We were eating better and using fewer resources--reducing our carbon footprint--but the reasons were more personal than political. So is the deeper question about why we discovered a latent gardener in our citified souls. That has no answer. The intangible rewards, the ones besides the vegetables themselves, were not something we were consciously aiming at. If it turned our lives around, it did it so slowly we barely noticed the motion. How can you barely notice planting eight hundred tomato seedlings? Of course I noticed how the summer routine changed radically over the first few years. But what I thought I was doing seemed the same, it just took more time. I suppose it's another example of the way we're so deluded by what we need to believe about ourselves, or what we lazily accept without reflection, that we don't know the truth. Perhaps because we started late in life after careers in television, the conversion to gardening felt oddly natural, the next thing to do. Not stopping to ask what we were doing was part of the naturalness. Like the famous Zen archer, we hit a target without aiming at it, without even knowing there was a target.

Gardening was part of my inheritance. Both my grandfathers grew up on farms and they and their families were also part of the great nineteenth century migration away from the farm, from working with your body to working with your brain. During their white-collar careers in town, they both kept small gardens. Like our friend Werner who grew up on a chicken farm they may have felt they should plant a small garden at least because that's what you did if you had a yard. People raised on a farm certainly didn't romanticize growing vegetables; it was more like a necessary part of ordinary life with no special virtues attached to it. For different reasons Kate and I seemed to have our own version of this practical attitude. We certainly weren't romantic. Gardening was work. The organic agriculture movement was all around us but it had a credulous aspect, as if it was based on a notion that if you just lived right with nature she would reward you with a productive garden, and beyond that with a healthy, happy life. I had no such confidence in nature. If nature had any bias toward virtuous human beings, I hadn't seen signs of it. (Maybe I wasn't virtuous enough.) There are enough disappointments for the gardener that exaggerating the virtues of agriculture and expecting it to infuse a vacant existence with purpose and significance seems silly. The weather can ruin a whole season's work with one storm. The sense of devastation can be wasting, and the feeling that nature has turned against you is depressing and nihilistic. Once you commit yourself to ambitious cultivation of the soil, you are stuck in one place, a fixed target committed to surviving the risks of being rooted in one spot. Instead of an enriching "life experience," it can be a blunt reminder of nature's indifference. Some members of the organic movement today--the newer word is "sustainability"--need to believe that nature is not just benign, she is positively beneficial if you know how to work with her. I think there is more than a little wishfulness in this attitude. If nature is helping you, she is also helping the fungus that causes blight, the squash bug, the hornworm, the slug, and all the other thoroughly organic pests that can share your garden. As the planet heats up due to us, it only gets harder to deal with the exceptional weather events that are becoming more frequent and more extreme. Looked at that way, the biggest pest in the garden is oneself, and one's fellow man.

10. The Marquesas

In Robert Louis Stevenson's *In the South Seas,* a book about life in the most idyllic part of the globe, he describes the shocked reaction of the local indigenous people when he told them their children would behave better if they were given a good beating. I think when he said "beating" he meant "spanking," but it shocked me too. He went on to describe how coddled the children were in Marquesan culture, that a child might stone his own mother without any fear of correction. That sounded extreme, but Stevenson meant it. To illustrate, he said if a child disliked an adult enough to say something about it to the other adults, that person's life might not be safe. Stevenson said he knew this because he was the beneficiary of the opposite behavior, that a child took a liking to him and as a result he was loaded with gifts by the child's family. They couldn't give him enough.

All this is a prologue to Stevenson's account of Polynesian attitudes created by see-sawing changes in population. At one point the islands had become overpopulated and this encouraged a policy of birth control, only two or just one child per couple. They even ate babies, or so it was said of them. They became known as cannibals. Then the Europeans came with smallpox, measles, and syphilis. Their missionaries also brought religion but it was disease that drastically upended their culture. Soon after controlling population by infanticide for fear of not being able to feed themselves, the population reversed course, plummeting so fast it looked to the natives as if they and their way of life might die out entirely. A policy of "unchastity" followed to encourage pregnancies, but it was too late and there was no vigor in it. Instead the drastic rate of mortality created a culture of death and the Marquesans lost their belief in life. "Fond as it may appear," says Stevenson, "we labor and refrain, not for the rewards of any single life, but with a timid eye upon the lives and memories of our successors; and where no one is to succeed, of his own family, or his own tongue, I doubt whether Rothschilds would make money or Cato practice virtue." This is Stevenson's interpretation of how people behave when so many of their own kind "and tongue" begin dropping dead. He notes that the Marquesans quit writing songs; the children who continued to perform them apologized for the scantiness of their repertoire but, they explained, they had lost interest in learning the songs, and no longer bothered to learn their own language either. The wan spirit of accidie had seeped into them and the whole island had begun breathing the air of futility. The death of everyone, of their entire culture, was looming unavoidably. They were doomed.

Reading this took me back to Eastwood, to Forest Hill Drive, to when I first felt my own fear of annihilation. It wasn't because people were dying all around me from disease. It was fear of the atomic bomb. At school we saw film of the burgeoning radioactive rings from the bombs dropped on Hiroshima and Nagasaki, the otherworldly clouds rising over the vaporized cities. Later we saw film of the bigger, better hydrogen bomb with its stately alien mushroom cloud thrusting upward with a similar but even greater swiftness and force. This was the ultimate man-made power. But it wasn't the ultimate. Out on the street, on tidy Forest Hill Drive where middle and working class people lived, I heard of the bomb that was even greater than the H-bomb. As much greater than the H-bomb as the hydrogen bomb was greater than the A-bomb. This one was called the C-bomb. Cobalt. It could blow up a whole country. Yes, the whole thing. A few of them could reduce the entire earth to a cinder. The kid who told us about this was someone I'd never seen in the neighborhood before. Perhaps that gave him more authority, but it was a boost he didn't need. My imagination was already stoked to believe the worst of these successive shocks. The A-bomb was something you couldn't imagine, yet it was real, it existed. Entire cities were blown up when it dropped. What would become the familiar experience of reality overwhelming anything you could imagine was initiated for us by the A-bomb. You didn't have to believe it existed because whether you did or not it was true. It was a fact, a new condition of life. So when a dirty faced kid with lollipop breath put his face in yours and said, "There's a bomb way bigger than the H-bomb, you know that? It could blow up the whole country!" there was no defense. We had learned the hierarchy of atomic bomb technology. First, one more brutal than anything you'd ever possibly heard of. Then, incredibly, an even bigger one. Finally, the ultimate, beyond description. They came in threes, from the horrific to the indescribable.

It didn't really matter that I never heard anyone else talk about the cobalt bomb besides the strange kid with the sickly sweet breath and a dirt-smeared face. I was primed to believe anything about nuclear annihilation. It was a new fact--like the gila monster, the dentist, the hypodermic needle --something else I had to live with. Unlike those however I couldn't isolate this one. It spread into everything, it was always there waiting, especially when I already felt weakened, threatened by something. Worse, the image could intrude into your mind when you felt happy and secure. There was no barrier you could put up in your mind that would block it. With the same unstoppable motion as the cloud rose in the air on the end of its mushroom stalk, the image of oblivion overspread whatever was there. Because it was man-made, it had nothing to do with God. He wasn't helpless against it, he was irrelevant. Man could do what he wanted and that included blowing everything up.

Perhaps God could prevent it but we didn't know how. The levers of this power were not in those omnipotent hands. He was more like a force of conscience, the hope we would use our freedom of choice to be better humans.

My fear of the Bomb was not an active one I had to squelch every day or it would have disabled me. The knowledge of complete annihilation was more like the weather. It was a constant aspect of life which I noticed according to circumstances and mood. On good days it was almost entirely absent. At those moments it was a rational fact that didn't stir any emotion. You couldn't laugh at it but a sardonic crack could keep it at bay. The worst times were those when something else reminded you of mortality, of human wishes for a long fruitful life. It was when something aroused the metaphysical organ. That was when you discovered that the organ had been taken over by the image of the swiftly rising, bigger than anything, all-annihilating mushroom cloud. For a fact, really, there might be no future. There was only this little respite from doom before some mad bully blew up mankind. All else was mere fondness, the weakest human force there was, so weak it barely registered.

This was how the spirit of accidie was introduced to me. It was not a vague sense that mortal existence was vulnerable, that organic life entailed risk, that nothing was truly guaranteed. Those sensations were different. This feeling I had learned from the fact and image of the atomic bomb was akin to what Stevenson was describing among the Maraquesans in 1888. I felt as vulnerable as they did on their scattered volcanic islands. They had nowhere to go to escape from the mysterious force killing their people, their culture, their language, their songs, killing their children and--if they survived to adulthood--killing their children's children too. That was the image of the eradicated future they carried in their heads, all of them--the blank space where the future should have been. Yet I was not on an island, I was not a Marquesan watching my friends and family die. There were times when it seemed as if something was wrong with me, that something had made me exaggerate this threat. Personalize it. Other people didn't seem to be dwelling on nuclear annihilation. Perhaps I wasn't cut out for earthly life.

11. Pyramid in space

If I imagine I'm a tourist standing by a pyramid near the Nile it would make the sight of another ancient tetrahedron out in space almost normal, instead of a whimsical absurdity. I might begin to see pyramids everywhere, as plentiful as stars! Yet to see just one gliding by Orion's belt or the Pleiades would upset everything I think I know about the cosmos. A whole lot else besides. But it's our little agreement, or conspiracy, to know it isn't true. We aren't like those addled or distraught witnesses who claim to have had aliens land in their backyard to take the family for a ride in space where they showed off the mother ship and the wristwatch you were wearing stopped keeping time. That's what you said, credulously, believing your watch could do that, "keep" time, which didn't mean keep it in place, stop it from moving, but keep track of it. To follow the motion of its hands as if they were signaling some kind of message to you. Maybe that time was running down and one day soon you would run out of this essential resource. Then what?

12. Pyramid in time

I had a perception about the dimension of time, that it exists only with motion. So, if a pyramid is stationary, it has no time. I'm not the only one who had that thought. Sir Walter Raleigh wrote, "Tell time it is but motion/Tell flesh it is but dust..." It's an intriguing conceit but Sir Walter and I are not physicists and his poem is probably better poetry than science. There may be some relationship between time and motion since anything in motion can be "timed" when it moves, although that might not be saying anything of interest.

The image of a pyramid out there in space, floating or hovering, existing, being in space, stands for something curious and provocative, besides its whimsical absurdity. Levitated, it becomes a personal pyramid, and very different from when it's an ancient historical relic sitting so fixed and inert by the Nile. Middle class tourists, Europeans and Americans and Asians, stroll around it, taking pictures of the pyramid's ancient mass. A stupendous fact from the past. The pharaohs' attempt at immortality. Another kind of time. We think of pyramids as emblematic of time because they are ancient. They've simply been here so long. Historical time is an aspect of no interest to physics which deals in light years, not mere centuries or millennia. During the few thousand years they've been by the Nile, the pyramids haven't moved. Or they haven't moved enough for us to notice it. They just sit and occasionally someone, in a guidebook for instance, describes them as "timeless." By that the writer means they've been there for many human generations stretching back into the dim and hazy past. They don't mean existing outside the dimension of time.

If a pyramid did appear suddenly in space, say orbiting the earth, we wouldn't describe it as "timeless." We don't really describe things in space as timeless, old as they are, perhaps because even though they're all in motion, they're so far from us they don't seem to move. Well, the sun and the moon seem to move, if you look away for a few minutes. All the rest seems as if it's in another dimension, and has been for millions or billions of years. (Which is a measure of time, of course.) Or so the cosmologists and astronomers say. Everything's orbiting something or flying away or approaching, nothing is static, so every object in space can be described by its dimension of time, its distance from something else.

To say time comes into existence with motion, that it's an emergent phenomenon, a product of movement, steps into the physicists' domain where the concepts are abstruse and highly nuanced and almost anything based on common sense is so wrong that, "it isn't even wrong," as Wolfgang Pauli, a physicist and a wit, said critiquing a colleague's theory. Almost anything one says about time is trite and obvious (time flies) and most likely pretentious because whatever time "really" is it seems to be everywhere present--or gone a moment ago--yet still here and simultaneously always anticipated, the silently ticking theme song of everyone's life (occasionally punctuated by laughter). Strangest of all to me is that despite time's omnipresence as a "fourth dimension," the physicists don't agree on what it is. Whether time is fundamental is not a settled question.

According to Lee Smolin, a theoretical physicist, Einstein's relativity theory implies a timeless universe. Smolin's own research and cogitations on time have taken him in the opposite direction. "Our experience of time's passage is the one thing we directly perceive about the world which is truly fundamental," he wrote in *Time Reborn*. "All the rest, including the impression that there are unchanging laws, is approximate and emergent." I feel a positive visceral response to the idea that time is real and can't be reversed. Having Dr. Smolin on my side is very encouraging. He has also concluded that there is only one universe and that the big bang was not the birth of the cosmos but an event in an ever evolving universe that exists in the same fundamental reality of time as we do. What we call the big bang may have been a new start for the universe arising from its evolution over time in a history of big bangs.

It is curious to put Smolin's ideas into my own spare words because they suddenly seem fictive themselves: how does he know things like this? Yet why do I think his ideas make sense when I'm reading them? The ones I understand, that is. For an amateur like me there is no way out of the problem of trying to understand ultimate existence by reading physics. Even when it is written by practitioners like Smolin and Stephen Hawking, it slips out of my mind almost as soon as I try to embrace it. I'm never going to experience the noetic insight I crave. I can only wonder enviously at what it's like inside their brains. In the meantime, down at my level, Smolin has given me two notions I find reassuring: that time is real both personally and physically, and there is only one universe. Again, when I say this so boldly, lifted from the pages of the book it was embedded in, it seems inadequate, too unsubtle to be true. But all these thoughts about the ultimate forms of existence are abstractions which will never coalesce into something as palpable as, say, a pyramid in space.

It was only a brief surprise to learn that as physics studies vaster and vaster spaces and ever more minuscule tranches of particles and waves, vanishing portions of mass and energy, it would find itself pulled back toward philosophy. This is fascinating since it wasn't so long ago that philosophy got booted out of science like a drunk bounced out of a bar. Now here it comes back in the front door, politely escorted by the same bouncer. What happened? If you read contemporary books on physics you are sometimes escorted yourself by the author into the newly respectable world of philosophy where philosophy's confidence in thinking abstractly can help explain how the world works. (At least temporarily, for the duration of reading, and also less true for Hawking than for Smolin, I should add.) For much of the twentieth century philosophy as a subject has been shrinking. It isn't the revered mental exercise it was back in ancient Greece, but the theoretical physicists like Dr. Smolin who want to explain to themselves and to us what the universe is like have felt the need to reach for this tool at times in their quests through quantum effects and the cosmos. What they are studying can never be experienced in any way but indirectly by a human being. No one can see an atom or a quark or a black hole. When physicists begin describing things so tiny or vast, the mind has nothing to hold on to and it's easy to become lost in a cloud of vaporous language. That's why they reach for math, an even stronger and more essential tool than philosophy.

I have to pause here to ask why anyone would want to read, without coercion, a book on physics. Well, Stephen Hawking's *A Brief History of Time* was a best seller and Lee Smolin has continued to publish book after book on contemporary physics and cosmology. Obviously there is someone out there besides me who wants to have these mysteries untangled even if all the attempt does is re-scramble their brain. In my case there may be another reason too. When I was young I was filled with Presbyterian theology and mythology, and in later years, when those notions lost the status of beliefs to become the shards of personal history, their demotion left a hole in my head. It was never going to be filled again since what was missing was absent because of an organic change, yet the loss left behind an appetite for metaphysical speculation, something physics has in abundance.

At some point it is hard not to read physics as if it is a creative product of the human mind, almost a work of fiction itself, as I think I almost said before, using the discoveries of the high-energy colliders and telescopes as texts. The authors--the physicists--studying this material don't have much freedom, they have to respect the many layered rules of the game, otherwise it's no longer physics, it really is fiction. But it can be hard not to feel that since all this is based on atomic or cosmic events one cannot see or detect except indirectly, that the scientist's creativity might be a significant part of his or her ability to describe to us a coherent world of invisible forces underlying the apparent world we live in. In other words, their imaginations are part of the process. I don't say this to discredit them. It's part of my attempt to wrap my mind around some of their insights. I don't doubt what they are doing. There have been many instances when discoveries by these priests of the invisible have been confirmed in real world applications so there is no question that they have some potent connections to reality. Some relationship to it. As they look ahead, probing with new theories, they often contradict each other. Smolin himself has called out some of his colleagues for weaving tales about, in particular, string theory, and challenged suggestions we live in a multiverse with an infinite number of copies of all of us. These ambitious notions can't be verified by observation or experiment and therefore, he says, the scientists who propose them are not really doing science. His standard is that what you propose must be falsifiable. Otherwise, you can say anything, and in that case you really might be creating fiction.

13. Dwindling

I always assumed that as you got older your body dwindled, your mind too, in a gradual downward arc. Now that I'm here I'm learning that being old isn't a continuous process that can be tracked as if one were gently gliding down a gradual, arithmetic slope. Maybe some faculties are like that, the kind Hugh monitored so compulsively with his stopwatch, measuring his speed, strength, and endurance, as if he were spying on his own organism, gathering evidence of its mortality (which did turn out to be conclusive). What I notice about the experience is that being old is more analogous to being young. Childhood is a specific domain with characteristics different from being grown up. Old is like that too. The contours of mind and attitude change in quality and nature with the shortened horizon. Anyone can say this is true, likely to be true, but when you experience it you realize that is just the point, that it's a particular and palpable experience, qualitatively different from being the adult you became, more or less, in your twenties, and remained, more or less, until...when? On one of those yesterdays, when you had already started becoming old. Each of us has his or her own physiological changes like insomnia, or waking up early, or irregular bowels, altered eyesight, or inability to digest alcohol comfortably, but the fundamental fact underlying all the physical, mental, and emotional alterations is that the planet has begun going on without you and in the not too distant future the subtraction of you from the population will happen in the same way the yesterdays began gliding by. Then this life you lived will be frozen between your very own numerical parentheses and nothing will change, ever again. That is so unlike anything else that has ever happened to you--except when it does happen it's already over, it happened, and what's left is the planet without you, and if another you had a longer view, like from a pyramid in space, you could see that every moment of your entire life was left behind in a way not so different from your final instant exit.

14. Lying

Some of the kids in the neighborhood could lie to your face. Terry was one of them. When I accused him of taking my bike, he said, "No I didn't." "Yes, you did." "I never touched it." I'd seen him take it and ride off. He left it down by the school. I wasn't the only one who saw him ride off on my bike. Freddy saw him too. I couldn't just let it go, I had to confront him. The way he lied amazed me. The way some kids lied always amazed me. I got scolded and even punished if I lied. I had some internal engine of shame which made me unable not to tell the truth unless I was desperate. What amazed me most was the way kids enjoyed lying. You could see it in their eyes. They challenged you to deny the lie. It was just their voice against yours so there was no way to win, to flatten their shamelessness and make them confess. Once they started lying they kept going, lying for the fun of it. It wasn't until we got the accident of the orange man in the White House that I realized why he and kids in the neighborhood enjoyed lying so much. It alters reality and makes the liar, the person making up the fiction, the owner of the altered reality, the prince of What Is. The orange man has got everyone around him lying too, like some of the TV broadcasters, all aiming to change their version of "reality" to fit his. These are the techniques refined in the twentieth century, the ones Orwell taught us to recognize. Here they are, alive again, in the person of "reporters" and "commentators" grinning while they lie on television, or indignant, assertive, just like Terry on Forest Hill Drive, denying he had swiped my bike. Then, to pay me back for confronting his lie with the truth, weak as it was, he scratched his name on the fender of my bike. He didn't try to take it again, but left his name so I'd know his mark was on it, that what he'd said was the way things were, and that any time he wanted my bike he could take it and there was nothing I could do to change it because it was really his. It said so right on the fender.

15. Blocks

The house was a typical bungalow on Peck Ave., a street of working class homes, many two-family. White with brown shutters, a shed dormer in front, another in back, a small yard with a driveway and garage. Just inside the front door, on the beige carpet, right in the center of the living room, was a layout of small wooden blocks with a glowing nimbus of enchantment around it. The blocks were carefully arranged to form neat spaces, both rectangular and odd-shaped irregular ones. The blocks themselves were different sizes, machine cut into squares and rectangles, some with an arch cut out, longer and shorter lengths, even a few triangular shapes, and mixed in were strays from another set of blocks with different proportions. There were also a number of human figures, plastic cowboys, Indians, and horses. The figures were bowlegged so their curved legs fit snugly on the saddles molded into the horse's running or standing figure. This construction he'd built, Philip told me, was a ranch, and when he said this I felt my imagination drawn out to embrace this fantasy world. It was the first of many versions--a fort, a castle, a spaceship--though we often returned to the western motif which was so familiar from early television and movies. After the first introduction to what he'd created, Philip invited me to build my own place. That established the pattern. We'd start by divvying up the blocks into two piles, strictly equal, then each of us constructed his own "set." Philip always provided the story, the running account of events which I echoed as I played out a similar tale beside him on the carpeted living room floor. His stories were always adventures shading into morality plays about good and bad guys, ending in the destruction of what we had each carefully laid out as a setting for the action. A good guy or two always survived, like Ishmael, so the promise of a new adventure was left for us to fulfill.

My mind went blank when I had to imagine what these plastic figures might be doing in our wood block ranches or castles, but Philip was facile in naming traits and motives, usually venal or criminal, and assigning actions to our pawns. I think that's what he enjoyed most; for me it was building the sets for these scenarios. I loved the feel of the blocks in my fingers, the way fitting several into a pattern could transform the shapes into their own world, and the sight of them laid out on the living room rug put me into a spell that was like nothing else. Soon I got my own set of blocks, mostly pieces of wood Dad helped me cut up from scraps in the basement. I added strays like dominoes and a few Lincoln logs from another store-bought kit and even a Tinker Toy or two from around the house. Then we could play "blocks," as we always called it, at my house as well as at Philip's. It was never as good at my house because my younger siblings always tried to play with us, hovering around, innocently interfering. The blocks I'd scavenged together weren't as invested with magic as Philip's were either. My collection wasn't neat and regular but a motley jumble of unmatched shapes, and the sets we could build with these materials had less coherence. More imagination was required to summon the magic that seemed to emerge from the constructions we made in Philip's living room. Our house wasn't a private haven like his where he was the only child, and the ruckus made by the younger kids affected the atmosphere we were trying to cultivate. Our shared narrative was already compromised because the living room was too public a space, the dining room had a big table and chairs in the way, and my bedroom was too small. Philip's house felt like a sanctuary centered on him and his activities, which included me when I was there. His grandmother sometimes sat in a corner of the living room and watched us, or dozed, but his mother and aunt were careful to avoid the living room once we spread out our blocks. Philip was articulate and precociously ironic instead of loud or rambunctious, and his family indulged him, often speaking to him like an adult. I absorbed this atmosphere readily, so different from the perpetual kindergarten at my house, and tried to act and speak the way Philip did, although I was always aware my imitation was feeble. The wry, mature way he had of expressing himself came from who he was and I could only admire it from outside. The name that Philip had given their cat seemed to sum up all this. When I asked what the tabby cat's name was, Philip said, "Said Cat." It was my introduction to the peculiar expression used when one wanted to refer to an object or person--or cat--previously mentioned, you said: *Said cat.* Philip not only knew this adult convention of speech but how to use it in his own talk.

Immersion in Philip's created world of blocks added an almost enchanted element to my own world, one which I could not have imagined myself. This created world, along with the other things we did together, all initiated by him, simultaneously concentrated and expanded my mental landscape. I hadn't known it was so cramped and barren until the powerful light of imagination from the blocks and our other activities lit it up. It was something like what drugs can do to a brain, a sharp peak of mental pleasure suddenly rising out of banal daily life and the numbing hours of school. I can't say playing blocks was what I'd been looking for because that part of my mind had been blank. I hadn't known what I was missing. But once I found it, I realized it was not only the most important thing, but to my surprise it was the very meaning of my life.

This thought first came to me walking home from school, under the maple trees, wondering how I was going to make it through the week in such a brain-dead mood. Then the thought spontaneously rose up: it was Wednesday and after only two more days of this it would be Saturday and I could call Philip and see if he wanted to get together. If he said yes the glow of anticipation swelled up and from that instant made me feel elevated. If something interfered, or Philip said he didn't feel like getting together (how could he *not*?), I felt bereft. What could I *do*?

Usually envying someone makes you uncomfortable, looking for a way to get even, but I felt free to envy Philip, even to show more esteem--more need--for him than I expected in return. The imbalance in our relationship was created by my wanting to be him, to have his life instead of mine, aware that he had no interest in being me. Strangely, he rarely if ever took advantage of this power. Perhaps he was aware that if he did we might revert to normal children's behavior and lose what was unique in each other's company. Whatever the reason, it was a delicate equilibrium of several forces, and quite adult.

Also contributing to the imbalance between us was that my family was normal and Philip's family was not: he had no father. His father, my uncle, had been killed in the war, in 1944, just a few months after Philip was born. It was not an event anyone talked about except Mom who had been close to her brother and liked to remember him. Now I can see that she was still trying to digest the fact of his loss just ten years before, still trying to understand what it meant for a young man to die leaving a baby and widow when his life has barely begun. Even if no one could alter this fact--and many others had died too--shouldn't they be doing something to make up for his loss? How could they all go on living normally? Yet what *could* anyone do? There *was* nothing to do except go on living normally, as if everything was normal, even if it wasn't. Yet somehow it wasn't right to act as if it was. This was the recycling conundrum Mom's mind twisted itself around when she thought about her lost brother. Her thoughts made him so alive it made her feel he might simply show up, walk in the door, as if nothing all that unusual had happened. This scene played itself out in her mind in other ways too, partially nurtured by her own father's expressed hope that maybe his son had taken refuge with a French family in the midst of combat after receiving a blow to the head or some other kind of concussion that jarred loose his knowledge of who he was. He'd be able to get along because he'd studied French for several years in college and could speak to the residents. Maybe it would take some time, even years, for him to recover his sense of identity. But someday, maybe, he would. Mom only hinted at these imagined scenes, and I suppose only spoke of them to me because I was so young. It was rare when any of the adults brought her brother's name into a conversation and then it was only in a mundane way. Nothing was ever said that would stir up emotion, or a sense of loss. Everyone took their cues from his widow, Philip's mother, who was very private. She was also much loved by everyone. This vacancy in the family illustrated and enhanced a message preached in church: life was a gift, it was sacred, a blessing from God. To keep that in mind, to live in that way, was also a means of honoring the one who had been lost. For reasons no one could fully understand, even if those reasons could be stated prosaically, his life had been taken away--either by God's design or by God's refusal to interfere with an incident of war. Or, I should say simply, his life was taken away because that's what happened in war; everyone knew that, so a stoic attitude of acceptance underlay everything. To complain would have been an offense both to God and man. The shadow of his father's absence fell over Philip's family's house, but instead of darkening it the shadow seemed to guard it, to shroud it in a protection everyone recognized and respected as a debt owed to those whose lives had been sacrificed for the country.

16. Saving the World

The view down Peck Ave. was the same as the one looking down Forest Hill Drive; it was the same view if you were standing on the plinth America had built for itself by "winning" World War Two, and now stood on so proudly beside Lady Liberty. "What am I doing here, holding this torch in my hand?" she asks. "Is this my reward--to stand here showing the way to the world forever?" Yet anyone who had read a few pages of history knew that if there was one country which had sacrificed the most for victory in that vast war it wasn't the U.S.A., it was the evil Russians, the monstrous U.S.S.R. But the Soviet Union's soldiers had mostly died out of our sight and as much as we'd needed them as an ally their country could not be our hero, especially not when we wanted the role for ourselves. Sort of. The stars and stripes unfurls with a thunderclap, I hear Tom Brokaw in his rumbling, self-intoxicating voice broadcasting the title of his book written to flatter our citizens, *The Greatest Generation*; then I hear Mom ask in a tentative voice, "Oh, is that what we were?" She is not sure that honor was worth her brother's life, if that's what it had to cost. Whomever Mom and Dad and their contemporaries thought they had become after having lived through the Great Depression and then World War Two, it was Beth and I and our cohort who were raised in the aura of America the Savior, not our parents. While we were learning to walk and talk, the Marshall Plan was spreading its wings over Europe and the high moral purpose this allowed the rhetoricians to celebrate (along with the realpolitik and capitalist goal of preventing Europe from "going Communist") infused us with a sense that we were here on earth to do something important and noble. Unprecedented too. The spirit flattered us, just as Brokaw's book flattered his audience half a century later, flattered them into playing unconscious collaborators by believing it. Given the forces in play, Beth and I were also helpless. The message burrowed straight into our secret heart while Kate Smith sang *God Bless America* on the black and white television Mom watched while she ironed clothes.

17. Miss Moore's gift

I notice something else too, looking down this view of Forest Hill Drive. The layout and architecture of the neighborhood manifested a solid, foursquare rectitude in its design. Down the middle of our modest street was the grassy median strip resembling a small park. It wasn't wide, twelve or fourteen feet, curbed all around and with two openings breaking up its length so cars didn't have to go all the way to the end of the block to reverse direction. These narrow passages turned the strip into a series of islands which also meant we could ride our bikes or roller skate around the elongated ovals, and it gave the whole street a recreational air, as if it had been designed for kids. At the end of our street, facing Sunnycrest Drive and at right angles to Forest Hill Drive, was Huntington Elementary School. Its long rectangular form was built of two storeys of bricks, capped by an ornamental concrete cornice. The right angles all these heavy construction materials were laid out in conveyed confidence in a solid regular world. The exterior esthetic effect echoed the foursquare self-assurance of the curriculum we were taught inside. The one graceful deviation from all the straight lines was the swooping curve of the sidewalk leading from the twin pairs of front doors down toward the street. Like Forest Hill Drive, the walk had a passageway cut between its two lanes, where kids could flow from one side to the other without stepping up over a curb. In the middle of this passageway was a flagpole with a flag flapping on top every day, completing the neatly designed setting for the familiar symbol of the country we lived in.

In Miss Moore's sixth grade class, toward which we had been advancing every year, we learned that the H-bomb was necessary because the Soviet Union had one too--stolen from us!--and their entire nation was constantly indoctrinated by communism with never a chance to learn about the freedom and democracy we enjoyed in America. Unlike the Soviet students, we were free to think as we wished and to hold our own opinions, a sacred right our forbears had fought for starting with the Revolution. Miss Moore's political philosophy agreed with William F. Buckley and the John Birch Society, and she fed us regular doses of it to counter the threat of communism. There were spies in our midst taking advantage of our freedom of speech to preach a godless creed, so the danger to the Republic was present all around us, every day. It was my first exposure to the paranoid strain in American politics. I believed what Miss Moore said about the lack of freedom in the USSR, her description of the Siberian labor camps, the penalties for criticizing the government, her portrait of the Iron Curtain separating the communist world from the free one, and most of all the threat of nuclear war now that the Russians had nuclear bombs too, an unspeakably nefarious development that had happened in our brief lifetimes because of those traitorous spies. It was our heritage to have to fight communism everywhere under the threat of nuclear annihilation. We had no choice but to live in a state of alertness and vigilance to protect the great principles of our Constitution from a godless communism. Thus the urgent arguments went, repeating the dire admonitions and menacing phrases. The dull, thudding sound they made in my ears could turn my fear of nuclear war into boredom, and I began to feel my well-being was menaced by Miss Moore herself in her black dress, hunched shoulders, dark hair knotted into a bun as she repeated her agitated warnings. At some point I also began wondering what it would be like to sit in a Soviet classroom listening to Miss Moore's Russian counterpart warn her students about the dangers of democracy. Soon I experienced an epiphany. Those students wouldn't know any more about what life on Forest Hill Drive was like than I did about communist Russia. I had been sufficiently trained by Miss Moore and the rest of our culture to bring down the shutters of taboo on this thought before it had infected my mind, but a glimpse was enough. A light had flashed behind the monolith. It didn't change my opinion, not right away, but it did plant a new darkly thrilling image in my mind. When I realized I was not supposed to think this way, that those thoughts were forbidden, the fact that they should be blocked became part of my picture of the Soviet Union. And since I was not supposed to consider any alternative to the truth

Miss Moore was urging on us, how could this be freedom of thought?

The insidious epiphany didn't change my opinion of evil Joe Stalin or the horrors of Siberian prison camps or people thrown in jail, beaten and tortured for questioning authority, or--as ever--fear of the annihilating mushroom cloud. But it did teach me the trick of turning my mind inside out so I could see the other side, the dark side, the side intentionally kept in the dark. I had felt a shudder that I was doing something strictly forbidden, something like committing a sin, doing something morally wrong. It was a mild sunny day in spring when this happened. I was walking along the sidewalk in front of the school and the thought made me stop, as if I didn't want motion to distract me from trying to understand what this mental flip really meant. The experience felt new and I couldn't remember having had such a subversive thought before. It was like opening a door and at your feet was a drop into nothing.

Despite the nice day, no one was out. I was alone on the street, under the sun. What if the inside out way of thinking was the only one you ever knew? If I were a kid in Russia would I suddenly wonder if everything they taught me in school was a lie? I crossed Sunnycrest and turned up Forest Hill Drive toward our house. If you never heard anything else but one version, how would you know? You wouldn't. Yet you could change everything just by the way you thought about it. It was an odd power to have; it couldn't really change reality although it could make it harder to see what that reality really was.

18. Peck Ave. Two

Peck Ave. was only several blocks away from Forest Hill Drive, but it was a different neighborhood, no longer in Eastwood, although the people who lived there were much the same--white, modest middle and working class, mostly in single-family houses but with a sprinkling of two-family ones mixed in. They were both Catholic and Protestant and a few might have identified themselves as Irish or Italian or Polish but the culture was homogenized American. After Philip moved away--taking our mesmerizing games with him--I never went back to Peck Ave., not until we moved to a house on the next street over from it. It was only a block but the houses on Rugby Road (a road not a street) were bigger with more spacious yards, edging into the upper middle-class. Mighty, venerable elms lined the gracefully curving chicane of Rugby Road, named by its developer to evoke leafy English country neighborhoods. The ambition for gentility was somewhat undermined by the zillions of children tumbling out of the oversized houses. The Rooneys had the most kids, nine or ten (we didn't know how many) who all went to parochial schools, so we didn't get to know them very well.

When I changed schools to the local junior high, I met Jimmy, and when he brought me home with him it took a long, time-distended moment to realize I had been on his street before--he lived on Peck Ave. Only Jimmy's Peck Ave. was nothing like Philip's. They were so different they were like two separate places, where I had two entirely different experiences, so different from each other I was now a different person too. I never tried to join them, to reconcile the two Peck Avenues into one. Instead I intentionally kept them separate and never even asked Jimmy if he had known my cousin. My intuition had already answered the question. Philip was different, more cultured and cerebral, he wasn't out on the street with the other kids who seemed characterized by his former neighbor Vito who had painted his dog blue with a can of paint and a brush.

Instead of a refuge, Jimmy was daily life, all vigor and high spirits. I saw him every day at school, we ran around the neighborhood together (though I never met Vito or his dog) and he soon became so familiar in our house he was like a member of the family. Jimmy would grow up to be a state policeman and if there was an ideal piece of human clay for that role, it was Jimmy. He was naturally fair, honest and decent, alert for good deeds to do, the kind who would help an old lady cross the street without being asked. He was also full of mischief and teasing, and might slip a whoopee cushion under you and hoot when you jumped.

19. Vote for Bobo

"Here," Jimmy said. "Put this on."

"What is it?"

"Put it on. It's a sign."

Jimmy was wearing one too, a cardboard sign tied with strings so it hung over his chest and back. Vote for Bobo, it said in bright crayoned letters.

"Who's Bobo?" I said. It was the first semester after we moved and I didn't know anyone besides Jimmy.

"He's running for something," Jimmy said. "He's a good guy."

"Bobo Gentile," I said, pronouncing it *jen*-tile.

"Gen-*tilly*," Jimmy corrected me. "He's in the ninth-grade." The school only had three grades and we were in the lowest. Later Jimmy pointed out a group of older guys in the hall and said, "That's Bobo," but I didn't know which one he meant. No one took any notice of us wearing signs. I was in some foreign country called Lincoln Junior High School and didn't know who it was wearing the sign but it wasn't me. I was afraid someone might ask me what I'd asked Jimmy, Who is Bobo? and I'd be unable to answer and get in trouble. Why are you wearing his name if you don't know who he is?

Lincoln had a long sloping corridor between the two old brick buildings that formed the school. All the halls had low ceilings, especially the cafeteria where we were packed in so tight it felt as if the walls would fall in on our waxed paper wrapped sandwiches and soup with blobs of fat floating in it with stewed tomatoes and rinds of something anonymous. The whole place was Dickensian with a threat of violence hanging over it. In art class when the teacher left the room for a moment, the kid behind Jimmy grabbed his hair and yanked his head back and held a razor to his neck, just to scare him. Or maybe it was only a joke. Jimmy didn't tell anyone besides me and if he had it might have led to more trouble. Mr. Sutton, the guidance counsellor, was friendly but with his short crewcut and sizing-up eyes he looked like a big cop packed into a suit, and it was known that if you got called into his office and he shut the door the guidance you received would be physical.

At the bottom of the hill from the strange world called simply "Lincoln," was one of the cushy neighborhoods in the city. Down the hill in the other direction was the "working class." So the school population was economically mixed. It was all white although the city was one-fifth black, and the most identifiable ethnic group was Italian with kids like the operatically named Giannina DiFuria and Rocco Fragola. Most of the other ethnic kids were a couple generations beyond Ellis Island so tensions that erupted in fights were usually personal. Billy LaBrake got picked on probably because he was a little overweight and clumsy. Mickey Vendetti who was actually a pretty good guy bloodied him one day after baiting him like a pit bull. Maybe there was something behind it but it looked gratuitous and ugly. Billy was so bland he seemed like no one in particular, a handy victim. The girls cooed Mickey's name and rolled their eyes because he was "cute" with a slick pompadour of black hair but he wasn't dangerous like Al Distiola who brought a gun to school. Al stopped Jimmy in the hall and showed it to him with that drugged look in his eyes that said he might use it on someone just for the hell of it. The music teacher Mr. Oberbrunner had made a project out of saving Al because he was a noticeably good trumpet player. I played trumpet too and sat next to him in orchestra but he never made a sign I was there. Or anyone else. He looked so dopey he must have been using some kind of drugs although this was quite early in the days of drugs in junior high schools. The only person I'd seen using anything was Tim Spoor sniffing varnish in wood shop till it made the pimples on his face run with pus. When Mr. Oberbrunner noticed Al had a natural gift for playing trumpet, he told him so in front of the rest of us, an attempt to praise him into responding since scolding would never do it. Mr. O had no authority besides persuasion since orchestra was voluntary. But Al was lost somewhere in his stoned, vacant mind, and only occasionally showed up for rehearsals. When he played, the sound that came out of his horn was something I admired and envied, a thing of beauty that everyone responded to except Al himself.

Tall and thin in his gray suit with his long heavy blond hair combed back crowning his head like a helmet and his glasses magnifying his sensitive, worried eyes, Mr. Oberbrunner was completely out of place at Lincoln. His voice was so soft and civilized you had to lean towards him to hear. Had he suddenly appeared on stage with a symphony orchestra playing "Afternoon of a Faun" on his flute--his primary instrument--he'd have suddenly looked right in place. If the Muses had a single devotee in the mediocre midst of Lincoln, it was Mr. Oberbrunner.

The line dividing the upper and lower social groups was evident but in a somewhat unconscious way; there were no gangs or named cliques and plenty of social mixing. I had good friends from both groups, including John Philips. John was older, having been held back a couple grades, and he was the biggest kid in the building. He already had a man's physique and walking the halls he looked like one of the staff or a teacher instead of a student. He was quiet but had a loud, horsey laugh like Andy Griffith in "No Time for Sergeants," playing the country boy who doesn't know his own strength. John lived around the corner from me and since our routes to school converged we often walked together. I had to take long strides, and jog a step now and then to keep up. I did most of the talking when we walked since talking for John seemed to require a physical effort. We were a classic pairing of the talkative, wisecracking undersized kid with the strong, gentle giant. Gentle, except that when he was aroused John became someone else. There was a vengeful grudge in his make-up, perhaps because his father had walked out on his mother and him some years ago so their life was a constant scramble for money to live on. In gym class gentle John would run over anyone to score, and Jimmy passed on the rumor that someone picked a fight with John and pulled a "Bowie knife" on him, which angered him so much he grabbed the knife and stabbed it into a telephone pole so deep the police couldn't get it out. Paul Bunyan stories like this followed John around as if we were on the frontier, a place he might have felt more at home than in an industrial city in the middle of what the next generation would call the rust belt. He never fit at school either and at the end of the year when he finished ninth grade he enlisted in the Marines, and was sent to Parris Island.

20. The bear in the room

A more complex display of physical contrasts was provided the day we were herded into the auditorium to see Jim Brown, the star running back for the Syracuse University football team. When he walked up the steps to the stage between Mr. Sutton and Mr. Kendrick, the principal, he seized our attention instantly. Jim Brown wasn't just big, he was bigger in every dimension than the statuesque, silver-haired principal. Despite his commanding size, he moved with a gracefulness that looked out of place wearing a suit and tie. Everything about him was different, unprecedented in our experience. There were no black students at Lincoln; they went to Roosevelt, and most of us had no "Negro" friends or even acquaintances. The word nigger was heard now and then at Lincoln and whatever else it meant besides color, it meant someone assumed to be inferior. Sitting in the auditorium looking up at Jim Brown in the flesh, in his black skin, we were looking at the very opposite of what the epithet implied. This man was unmistakably superior, a godlike human being, more physically imposing than anyone we'd ever seen before. When he addressed us from the stage, his manner was a mixture of reticence and dignity, yet how could he not be aware of the oddness of the situation, even more aware than we were. A lone Black man standing in front of several hundred adolescent white kids telling them to work hard and use their abilities so they might succeed. Succeed in a society where Jim Brown's own role, despite his unique athletic gifts, was defined and restricted in a way that ours were not, as puny and common as our talents might be.

This anomaly meant the message we got from "meeting" Jim Brown was mixed and confusing. He had not been invited to Lincoln to start a discussion about race in America, but race was the bear up on the stage beside him. More than beside him--the bear *was* him, the bear of his history, and our country's long bloody history. His ancestors had been slaves, yet who in that auditorium would have had the strength--the balls!--to enslave a man like Jim Brown? We teenagers, especially the male ones, knew the answer to that. Only by some inversion of nature or perversion in the arrangements of society, could we dominate such a man. By putting this fact, this contrast, in front of us, our country's upside down history was displayed while we listened to the platitudes recited by Mr. Kendricks and Mr. Sutton and by Mr. Brown himself.

What the event meant for us students more prosaically was that we got to skip a class to "meet" a famous athlete. But the deeper message was one we all carried back to class with us in some form, a message that Jim Brown must have borne much more heavily and habitually. Our superiority was merely in numbers and the social and quasi-legal prerogative of whiteness, but his was individual, and for the blessing and burden of his so-called God-given talent he was at times resented, even hated, as much as admired and envied. The power contained in his person could be scary and for some it was a comfort to see it on show only when he was wearing a bright orange jersey and running around a football field on a Saturday afternoon, carrying nothing more threatening than a grenade shaped leather ball while trying to win a game for "us," his fans.

21. The lab

Jimmy had the power to banish the fog of accidie, or distract me from it so I could rejoin the life around me. He was naturally upbeat and cheerful, and if I mentioned the Bomb and the headache it gave me at times, he would have made a crack or said nothing. But I didn't mention it, not to him. I was aware of my audience. It wasn't something I talked about, not until later. Whatever he thought about it himself, Jimmy didn't say. He didn't talk about things like the Bomb. He didn't complain. He was a normal kid with a lot of energy and an unusually resourceful spirit. If he ever had a bad day, that too was private. It was the atmosphere at his house. You could not complain around his father. One look at Mr. Meadows and you knew the purpose of life was to work. He didn't work maniacally, but steadily, constantly. As a fireman he was assigned to rotating shifts so when he was on the night shift he drove to the fire station and went to bed. If there were no alarms he might get a decent night's sleep. Those weeks he came home in the morning and went to a second job unloading trucks in a warehouse or driving a liquor truck. When Mr. Meadows wasn't working he slept, or ate, or did household chores, or washed his car. Sometimes he took his wife on errands since she didn't drive. He watched a television show once or twice a week but no more. Mrs. Meadows was as chatty as her husband was taciturn. You could see where Jimmy's personality came from. Maybe his older brother Howie's too since he was edgy, a needler, and Mrs. Meadows despite her warm, amused eyes could tease or make a barbed comment too.

Jimmy was always in motion, and his mind was eager too, not intellectual, but curious and always ready to try something new. I had set up a lab in the basement of our house, an obvious imitation of the place Dad went to work every day, although there wasn't much in the "lab" except my chemistry set, a folding metal case with racks for test tubes and samples of chemicals. The experiments you could do with it were simple and harmless. If you mixed two chemicals in a beaker of water, the combination reacted together so a new compound precipitated out, making the clear water cloudy before it settled to the bottom of the beaker. That was too tame. Jimmy wanted to blow something up. I was game for that too although I also had my usual cathectic relationship to prized objects like the chemistry set, and wanted to preserve it in a pristine state. How useless! Jimmy hustled me out of this inert stasis but there wasn't much we could do that satisfied the urge for sensation. We didn't want to use an ordinary explosive like a firecracker--that was too easy. We wanted to do something that would reveal the powerful forces latent in chemicals, a power like the genie Aladdin released by rubbing the lamp, for instance. Something dramatic. We puttered about, not really doing anything, even after we rode our bikes out to the edge of the city to a chemical supply store which sold laboratory glassware. It was exciting to learn there was a place like this, and I bought a couple beakers and Erlenmeyer flasks not knowing what I'd do with them. In the absence of purpose or direction, my lab needed more props, and when I added the new glassware to some used implements Dad had brought home from work, including a genuine test tube rack stained with chemicals and a metal stand with clamps to hold a flask, it began to feel as if I had a real lab. Then we tried to set up another lab in the attic of Jimmy's house for no good reason except to give him a "lab" of his own too. I was feeling self-conscious about my privileged position compared to Jimmy so was glad to help him. We cleared a space, set up a small table, and mostly sat up there listening to Jimmy's portable radio and imagining what we could do if we only had a Bunsen burner. When Jimmy asked his father if we could use a candle to heat liquid in a beaker the phlegmatic Mr. Meadows became almost apoplectic. What were we thinking? He was a fireman! "Do you know how many houses I've seen set on fire by candles?" he growled. "Don't play with fire, Jimmy--it's not a toy!" A toy! This showed how pathetic our "labs" were. We didn't even have an alcohol lamp, and now candles were banned. Our dream of a real Bunsen burner like the one shown in a catalog I'd managed to get my hands on was dead. My father said no to that fantasy too, almost as vehemently as Mr. Meadows. All this puttering with chemistry was feckless. Then Donald invited us to come over to his house to see his lab. Donald was a half-grade behind us at school, but he was smart and disciplined. When he got interested in something new he studied it. Compared to him Jimmy and I were dilettantes. Donald's lab was a table in a relatively spacious, clean and uncluttered cellar. His father traveled for work and didn't always know what was going on at home when he was on the road so Donald didn't have a fireman or a chemist monitoring his lab. Along with a small but impressive array of equipment Donald had an alcohol lamp. He also had some potent chemicals. He showed us a cork-stoppered glass jar half-filled with dark purple crystals. "Potassium permanganate," he said knowingly, as he held it up to the light. "If you mix this with a little sulfuric acid..." He looked at Jimmy and me as if we knew.

"Let's try it," Jimmy said.

Donald hesitated a moment. He tapped out a layer of the deep purple, jagged crystals into a round-bottomed flask and clamped it on the stand. The acid was in another wondrously exotic and authentic looking jar. It had a heavy glass stopper and Donald diabolically invited us to sniff it. When we tried, he pulled it away quickly. "Are you *crazy*?" He lit the alcohol lamp and slid it under the flask. Then he carefully decanted a couple ounces of the acid. Almost instantly, the permanganate crystals dissolved and a layer of thick purple vapor formed over it. We all stepped back from the flask. The vapor had barely reached the neck of the flask when there was a loud BOOM and Donald, standing closer than Jimmy and me, recoiled and grabbed his belly with both hands. "Oww!" he shouted, more surprised than hurt, and we stood there looking at the neck of the flask still clamped to the stand. The bottom was gone. Miraculously the glass shards hadn't hurt any of us. Most of the flask had hit Donald's midriff but he was protected by his heavy flannel shirt and now he had his hands under the faucet of the laundry tub, washing off the acid and slivers of glass that hadn't stuck in his shirt. He was taking off his shirt when we heard his mother's voice from the top of the cellar stairs. "Donald? Did I hear something? Are you all right down there?"

Jimmy moved toward her voice and shouted up the stairs, "It's okay. We just dropped a vacuum!"

"Well it sounded like an explosion. Don't you boys do anything dangerous down there!"

The three of us were standing there, shocked, not looking at each other.

"I shouldn't have heated it," Donald said, sounding authoritative and sheepish at the same time. "That was stupid. The permanganate oxidizes very easily."

22. Firemen

Jimmy celebrated his own eagerness and his instinct for thinking outside the box. His cheerfulness somehow seemed part of it, a natural ebullience and refusal to accept the familiar, the drab, the known. I noticed how different his father was from Jimmy because his father's demeanor scared me. Mr. Meadows had no sense of humor, or never displayed any in his own home, at least not with me there. He always looked as if he was glowering at something. His manner seemed to go with the character and gravity of his job. As a fireman, he'd had people's lives in his hands, and would again too.

Mr. Meadows' code of behavior, his sense of values, his righteousness, blossomed vocally in his eldest son, Howie. Three years older than Jimmy, Howie was the extroverted version of their father and all his rarely expressed attitudes and opinions. Instead of broad and thick like his Dad, he was lanky and angular, and had a mouthy, mercurial edge unlike Jimmy's friendliness. Mr. Meadows had joined the fire department, he told his sons, because he needed a job during the Depression, but for Howie fighting fires was a mission. Saving people and their homes was the most heroic and worthy thing you could do and he grew up with the image of himself as a fireman as if it was projected on a screen in front of him where he could constantly see it. As soon as he turned twenty-one, he could take the exam and start training. In Howie's small bedroom--a converted porch off the living room--there was a shortwave radio permanently switched on and tuned to the frequency used by the Syracuse Fire Department. The volume was turned up so if there was a call for one of the companies--a "company" Jimmy quickly taught me was any single firefighting vehicle--the dispatcher's scratchy, staticky voice was broadcast through the house telling the firemen the address and nature of the emergency. If Howie was home and the call was for a company to go fight a fire, he responded just as if he was a working fireman at a station. He grabbed his clothes and car keys, ran down the staircase to the driveway, and jumped in his fire-engine red (what else?) Chevy convertible, backed out to the street, and raced to the address of the fire with the goal of getting there first, before any official truck or vehicle arrived. One day at the house not long after I'd met Jimmy, he stopped mid-sentence to listen to the radio, shouted, "C'mon!" and pulled me down the stairs. We could already hear the Chevy start up by the garage and right behind us Mrs. Meadows was moving as fast as she could, yelling, "How-eeeee!! Wait for meeee!" The staircase emerged from the house midway down the driveway and for some reason--maybe because his mother was shouting--Howie actually stopped at the door and we all piled in, Mom in the front passenger seat, Jimmy and me in back. I don't know why Howie didn't have so many speeding tickets his license was revoked because he took the race to the fire seriously. He stopped for red lights, but barely, and then if there was no traffic crossing the intersection, he floored it again. The pulse of its modest six cylinder mill was enhanced by a Glas-Pac muffler that smoothed out and shaped but barely muted the violent sound of an internal combustion engine. It had a white "rag-top" for bad weather and was painted the reddest red you could buy. Mrs. Meadows' constant screaming at Howie to slow down made it seem like we were going faster than we were. Although Howie was driving too fast he was not really reckless except for failing to wait out the red lights. The call was to a building on the other side of the city but Howie had heard the fire radio when the call first came in, so he had a good start, and knew the city's streets like a veteran cabdriver.

After all the excitement the neighborhood was weirdly quiet when he pulled up near a two-family house with smoke rising behind it and no fire apparatus in sight. Howie was out and running across the street as a siren sounded a few blocks away and a moment later a pumper arrived. I'd learned from Jimmy that the firemen called a hook-and-ladder simply a "truck" and one of these came around the corner with its long, articulated hind section swinging way out against oncoming traffic to make the turn, steered by the fireman perched way at the end between the ladders and eerily illuminated by its own flashing red light. It was a real fire! A police car pulled up behind Howie's unofficial flaming-red convertible, and the whole area was taken over by men in rubber coats, helmets and boots, unspooling hoses and cordoning off the area smoke was rising from. We couldn't see flames but all the firemen were in action and Howie had disappeared in their midst. The city firemen all knew his brother, Jimmy explained, and let him help almost as if he was one of them. I couldn't quite believe they'd let him handle a hose or climb a ladder but he had certainly plunged in among the real firemen. We sat in the car and watched, trying to look as if we belonged there. Howie had parked so close that we were behind the police line but the cops ignored us and I wondered if they knew Howie too, just like the firemen, and that's why he hadn't lost his driver's license.

Sitting in the red Chevy on the wrong side of the police line watching the scene, we were enjoying a privileged position and it seemed to stun us into silence. From our point of view, Howie was the director of this action, and we were waiting to see what he did, or told us to do next. In the meantime we sat, watching as if that was our job. What I was thinking about, quite unaware of it, was the role of the hero. In my youthful thinking, there was no actual place for Howie's behavior or how it seemed to be tolerated. There was no official attitude expressed by anyone; he was simply getting away with playing a fireman. Unconsciously, I saw the hero as someone who fulfills a role prepared for him, not one who creates it by breaking the rules. So this was a lesson, even if it hadn't bubbled up to the surface yet. It didn't even occur to me that if Jimmy had told me about this incident without my being here I would have thought he was exaggerating. Because it couldn't really have happened. How could mere enthusiasm for his father's job let Howie get away with this?

When Howie finally came back toward his car, his head down and shoulders hunched as if he and the others had finished their job, the firemen were packing up their gear. One of the "companies" had already left for the station and the hook-and-ladder began maneuvering its way toward the main street while the police directed traffic. Howie jumped in the driver's seat without a word. He tolerated a question from his mother and said simply, "Kitchen fire." After she asked a second time, he said, "No, no one hurt." We rode home in the gathering dark to the chesty rumble of the Glas-Pac muffler, still going too fast but stopping and waiting for traffic signals to turn green instead of just tapping the brakes to slow for red, then powering through.

23. Beriberi

My lab was really just a private place to get away from adults and siblings. What I enjoyed was setting it up. Once done, there was nothing to do besides trying those tame "experiments" in my chemistry set booklet. What I learned doing a real experiment was how tedious it was. It probably shouldn't be called a real experiment since there was nothing uncertain or original about it. I "proved" if you deprived rats of thiamine they got beriberi. That was already well known; it was true for humans too. The cruelty to the six rats who were my subjects was the other lesson I learned. They lived in #10 cans and a full grown rat was about as long as the diameter of the can, so once in the can the rat could turn around and stretch up to drink from the spout of the water bottle hanging above, but that was the limit of its motion. The cans were set on a sheet of wire mesh screen mounted on a frame of 2x4s, so their droppings fell through the screen to the sheet of newspaper below. Every day when I came home from school I had to weigh each rat, weigh the food so they each got the same amount, fill their water bottles, and change the newspaper. The ground meal they ate was identical for both the three control group rats and the three unlucky ones who were being deprived of vitamin B1, thiamine, as the point of the experiment. When we mixed up the two batches of food in Dad's lab in the research building at Bristol Laboratories, we added what looked like barely a pinch of white powder to one of the brown gallon jars. We had carefully weighed it out on a set of sensitive scales; I sprinkled it in from the metal saucer, and mixed and mixed and mixed it into the meal with a long paddle. It was impressive that such a small amount of a critical nutrient was the only difference in the diets of the two groups and if there was anything else learned by doing the experiment, besides the process itself, it was the graphic demonstration of how few grams of a critical compound could be the difference between normal health and crippling disease. In just a few weeks I saw results in the growth of the young rats.

To weigh them I had to open the wire mesh top, reach in and pick up the rat, and put it in another can on the scales to weigh it, subtracting the weight of the empty can, of course. At first the rats tolerated this. But as they grew and got furiously bored in their tight, uncomfortable homes--which seemed to shrink every day as they grew--they began biting the hand that came down daily to pick them up. I started wearing leather gloves but as the rats matured their teeth sometimes penetrated the glove. Dad saw me hovering over one of the cages hesitantly and said, "What's wrong?"

"They've started biting," I said. "Especially the deprived ones."

"Here," he said. "Just reach in and pick them up."

He put his bare hand down in the can and almost in the same motion his hand jerked up with a rat hanging on his finger by its teeth. A jet of bright red blood was already running down his finger.

"I told you, it's not that easy," I said, embarrassed for him but also glad he'd seen first hand how much trouble I was having catching the rats without getting bit. "Try the glove," I said helpfully.

By now the three rats deprived of thiamine were thin, their spines crippled from the effects of beriberi, the disease that results from thiamine deprivation in both rats and humans. The healthy ones had bright eyes, pink noses, sleek white coats, and had gained weight normally, as the scales confirmed. What they didn't have was space. Confinement in the #10 can was turning their originally friendly temperaments into hostility. The tedium of routine was affecting me too and I dreaded the daily afternoon feeding and weighing, noting all the numbers, and avoiding their teeth. The sick ones were depressing to look at too, and Dad agreed I'd carried the experiment on long enough to declare it a success and end it. The next morning when I looked in the cages one of the sickened rats was dead. Immediately I began giving them all the food with thiamine in it and it was gratifying and also instructive to see how quickly the sick, crippled rats began putting on weight and regaining some mobility. I dutifully drew a graph on a large sheet of cardboard showing the difference between the growth rates of the properly fed rats and the thiamine-deprived ones, wrote up my results and delivered my presentation at the science fair. I thought it was only my slightly hammy, mildly humorous performance that won second place for my grade and even that I felt was undeserved. All the older kids did experiments that I thought were so much more impressive and creatively conceived, with several complex steps before they reached a conclusion. I couldn't imagine doing anything similar myself. My modesty irritated Mr. Plume who saw my ribbon as praise for his teaching. He said the judge told him I'd have won first place if in the interview I hadn't kept saying, "My father helped me..." in describing each step of the experiment. Well, that was the truth. I'd done the drudgery but Dad had guided me the whole way. Besides the unintended lessons, I'd also learned that an experiment begins with a question, then is executed in an exact process of tedium and repetition, only slightly relieved by the reward of observation. Except in my case it meant watching three healthy rats get sick, so I also learned that the practice of science can be heartless. All I had done was demonstrate the well known fact that thiamine deficiency leads to a fatal disease, and my collaborators were the sacrificed rats. It was hard to say how their lives as my subjects were different from torture. Yet I still believed science should do what it did and use animals as part of experiments, when necessary, which was much of the time. The enemy was disease and it had no mercy either. To thwart all the ailments that could afflict human beings required making war on them and anyone who gave quarter was not going to win. When the experiment and presentation were all done, however, I couldn't help feeling as if I was a coward. I had done the orthodox experiment instead of standing up for the rats, which would have been a very different kind of experiment, examining human nature and morality, not biochemistry. So instead of proud of myself and my second place ribbon, I felt vaguely ashamed. I didn't like to think or talk about what I'd done. When Mr. Plume urged me to repeat my presentation for his science class at school, I got through it by making jokes and wisecracks, trying to make the guys laugh and the girls smile. But it wasn't funny or charming and felt as if the rats were suffering all over again.

24. God's country

Dick took his coffee black and smoked Pall Malls. He was in his twenties, a tall, wavy haired guy who hunched over his coffee as if he was protecting it from something, or maybe it was his privacy he was guarding. Yet he was always friendly and knew everyone, an Eastwood guy. Despite his apparent normalcy Dick had an abstracted air at times and a private laugh, a snort with a headshake, that made you think his world only partly overlapped with everyone else's. He sometimes laughed at John Betts' behavior, the way us kids--the soda jerks who worked behind the fountain--did privately. "God's country," Dick echoed with amusement and his distinctive headshake after Mr. Betts pronounced the verdict when another customer complained about the weather. It was a typical chilly gray day and everyone was ready and aching for spring. "God's country, Dick," Mr. Betts repeated back to him again and Dick smiled at him over his shoulder as Mr. Betts ducked back into the apothecary.

"It's the best of all possible worlds," I said to Dick from behind the fountain. It was a phrase I'd just read, and had been wanting to try out on someone.

"Candide!" Dick exclaimed, eyes widening.

We had surprised each other. We may have been living in God's country but neither of us expected to hear a line from Voltaire spoken in a drugstore in Eastwood.

"Where'd you get that?" Dick asked, shaking his head now in a parody of surprise. I told him I'd been reading Voltaire's book and we talked about it for a few minutes, unimpeded by self-consciousness since there was no one else at the counter. I knew nothing of Dick's life outside the drugstore but it was apparent he had more than the superficial acquaintance with Voltaire I had from hearing some guys at school talk about his famous satire, and that the book was a key piece of his mental experience.

"Our world would be different today if this guy hadn't lived," Dick said. "You know Voltaire wrote that book after an earthquake in Lisbon killed forty thousand people--forty *thousand*! They started asking their priests, 'Why did God let it happen?' What could the priests say?"

Dick said it as if he had heard his own priest's answer. He crushed the butt of his Pall Mall in the small glass ashtray on the counter and stood up to full height with his familiar heron-like unfolding and stretching motion.

"What could they say?" he repeated.

The front door of the store opened and a young woman came in leading a toddler by the hand.

"God's country," Dick winked and sauntered past the young woman and out the front door into the overcast late winter day.

25. Forty-eight

My ambition was to visit each of the forty-eight states. I wanted to be able to say I had put my foot on the ground in each one of them. I don't know where this idea came from because it had nothing to do with scenery or culture or history or any geographic characteristics of the states. They were names on the map, divided by lines, some straight and some wiggly, big and small, irregular and dissimilar except for the squares, Colorado and Wyoming, and the rectangular Dakotas plus the one missing a corner, Utah, and New Mexico and Arizona, near-squares, and Nebraska, a chunk out of its jaw and its back shaved off so it looked like a beaver. There were a few other odd pairings like Alabama and Mississippi, two halves of an open book (you don't want to read it), and Tennessee and Kentucky, both long and slender, lying down, although Kentucky looked like a club with the knout at one end, and Maine up there like the Statue of Liberty's torch matched by the other peninsula at the bottom right of the country, Florida, which looked as if some of the land in Georgia had oozed out into the ocean and improbably frozen there. And Texas, not resembling anything else and way too big as if one of the states had metastasized into a shape like a ship with a smokestack and keel--or was it a plowshare?--going somewhere and to hell with the rest of us. And California, cradling the weight of the truncated trapezoid Nevada in its lap. If you tipped the whole thing back as if California was the seat or fundament (mocking the usual joke about California being the place all the loose marbles roll down to), then Maine and Florida would look like horns at either end of the top. So the shapes, which you couldn't see on the ground, of course, were curious themselves, even if their characteristics were often arbitrary instead of natural, or freaks of historical events that had solidified into facts.

To be in each state, to cross the border into it, was like collecting stamps or baseball cards, and as oddly satisfying as it was to acquire each new state, to take possession by stamping a foot in it, the collection counted more than the individual states. They were totems in the larger abstraction called the United States of America, illustrated by the familiar map that hung in every classroom. Once the name had sounded so improvised, so made up, that it could never do as the name for a single, plausible nation, although it could be argued there was an analogy in "The United Kingdom" or "The Holy Roman Empire." I had no interest in these inflated names or the historical artifacts they referred to. At the same time, I couldn't say what it was about this abstract collection of forty-eight states that had captured my attention.

As a collector, it felt as if I was merely greedy. Anyone who collects things recognizes the covetous nature of their desire. Competitive, too. I didn't want anyone else collecting these states. They were mine. Or would be as soon as I had put my foot on them, or in them. But as real as the states are as geographic locations, real estate, the ground we all walk on, they have a characteristic the Scottish philosopher Francis Hutcheson noticed. One person's appreciation of something does not prevent another's. We can both get esthetic pleasure from a flower, a sunset, a work of art, without taking anything from the object or from anyone else. We can stand side by side and appreciate it together. "Consuming" it does not devour or lessen its beauty. (He went on to apply this to religious experience, saying the spirit was unbounded and inexhaustible.) But my collection was not something I shared with anyone else; instead it felt like something distinctly mine. Yet I couldn't say what it meant to me or give a name to the peculiarly elevated sensation it could at times inspire--the feeling I got when I rolled over the states in my mind, admiring my collection.

Regarded more practically, collecting the states was a way to mark events, although they were often disappointingly mundane. First was Ohio, shaped like a badge or a shield, where I was born, and second was New York where we soon moved. Because they were home they had nothing exotic or special about them and were taken for granted. Everyone had to be somewhere and that's where I was. In Pennsylvania there were relatives, a not very long drive away, so that state was soon acquired, and I was surprised to learn one day that I had been through West Virginia on one of our journeys from Ohio. There was a sliver of it that stood up between Ohio and Pennsylvania, something I hadn't noticed before Mom pointed it out. So I hadn't been in the main bulk of the state but just the narrow peninsula squeezed between the boundaries of two other states. It was pleasing to learn about this gift but if I hadn't even known I was in West Virginia, how could I include it in the collection? This raised an existential question, one that didn't seem to have a clear answer. If I were competing with another kid over how many states each of us had been to, I would have included West Virginia, no question about it; the game was mere numbers and I wanted to win. But in the privacy of my own mind, West Virginia was set apart in a special category. I might claim it publicly, just for the record, but if I had no memory of having been there I knew it didn't really count. I couldn't exactly define my collection of states but one thing I knew for sure was that it only existed in my conscious mind. To include a state, I had to feel--I had to know--I had really been there. The tests were consciousness and memory, even if I couldn't have said so myself at the time. Not that definitively, anyway. Placing a foot on the ground had to have this mental counterpart which I was aware of and remembered to make it real.

Which also raised the question again--why was I collecting the states?

On family trips we drove through and camped in the heart of New England--Vermont, New Hampshire, and Massachusetts. The others--Connecticut and Rhode Island--seemed close enough that they would be mine in the course of ordinary events, without any great effort. The exception was Maine which stuck out like the thumb it resembled and was going to be harder to get. It might require its own special trip. We also went north across the St. Lawrence River into Ontario which wasn't a state but a province of Canada, a foreign country. Fine, I'd accept it, but it gave me no thrill. It might have been more of New York State for all it added to my collection. The mental conception I was trying to fulfill was the forty-eight states of continental America and nothing north or south of the borders counted or mattered. My goal was perfectly provincial, to collect the USA.

In 1954 when I was ten years old, I took a trip with Philip and his family that added a hatful of names to my list. They were colorful, exotic, distant states which besides being new acquisitions also looked much different from anything I had seen in the east. Driving west beyond Ohio, the country spread and flattened out into a rolling landscape, the endless horizons of Indiana, Illinois, Iowa, Missouri, and Kansas, and then we were in Colorado. It was so unpredictably different in another new way, with its natural stonehenges of red rock thrusting into preternaturally superclear air, it felt like we must be on another planet. What a wonderful surprise! Acquiring this one state put a glow on the whole collection the way one treasured possession, a Willie Mays or Ted Williams baseball card, can dignify a whole stack of nearly anonymous players. Wyoming was almost as good a prize, especially with its musical name conjuring the image of cowboys (there was one on the license plate!) , and on the return trip we drove through South Dakota, with the Badlands and Mt. Rushmore, then Minnesota and Wisconsin. Despite this bounty, which by the end back in Chicago had overwhelmed my imagination with too much that was new to digest, there was one missed chance that rankled me. In Yellowstone Park, in the northwest corner of Wyoming, I looked at the map and saw that we were a mere ten miles from Montana. Yes, it was twenty round trip, but it would only take a half hour or so to drive to the sign announcing we were entering Montana, get out and put both feet on the ground, then get back in and return. It would be the first (and only) time I entered a state for no reason except to acquire it, which would blemish the achievement by having no primary purpose for going there, not part of the course of events. I didn't care, I was greedy, and "Montana" was an even better name than "Wyoming." My aunt, the driver on the trip along with her sister, said no. She said it gently because she was a kind person, much too kind to argue with, but she also said it as if the request was a fly she was vaguely swatting away. "Don't you want to be able to say you've been to Montana?" I asked trying to sound teasing instead of petulant, but the conversation had moved elsewhere. Even my cousin didn't care. That really surprised me because we were so close. Didn't he feel a compulsion to touch down in each of the states? We'd never talked about it but I realized I had assumed that this ambition, this strange greed, was probably something else we shared even if it might not be universal. No, apparently not. Philip didn't care whether he could say he had "been to Montana." He had not been collecting states as we crossed the country but merely passing through them. Actual places or sites might interest him but the abstraction of geographical-political entities had no special meaning for him. This discovery made me feel silly about my collection. What difference did it make if someone had "been to Montana" or not?

Besides adding many faraway names to my collection, this trip was the first time I became so distinctly aware of my peculiar ambition, and learned that not everyone shared it. When Mom asked me if I wanted to take a trip to Colorado with my cousin, part of the thrill was the thought of adding so many new states to the collection. Distant ones, too, the kind that were hard to acquire. I still felt the same way, silly or not, and would never again take such a bountiful trip collecting states.

The psychologists say collecting is a way of controlling, even if only symbolically, but I think of it as a way of giving meaning to things by making them your own in some way. By saying I had been in a state I could attach it to myself and say it was mine. I'd acquired it. It could belong to anyone else too, in the Hutcheson sense of sharing the intangible. My claim wasn't exclusive, only my collection was, the same way people collect birds for their life-list. There was something else about this peculiar ambition which makes me think that it springs from some tribal urge and may be, at bottom, universal. When we were introduced to the concept of Manifest Destiny in school, I understood it in some relation to this collection of forty-eight states I was amassing. I should say I felt a relationship, not understood one, because the connection was emotional. Viewed through "Manifest Destiny" it looked as if I wasn't the only one who'd felt this lust to collect sections of land marked out on a map and keep them in a purely abstract container, one which I identified with my own physical being. My childish fantasy had overlapped the real world. "Manifest Destiny" may have been a political slogan expressing a popular form of the urge I felt while gathering my abstract collection, but the lust for land it represented was real enough. It was wealth and power, it was empire, it was blood and crime--it was, at times, war. Manifest Destiny was a hunger and fed to its limit it would have acquired Canada and Mexico to include the whole of North America as part of the United States. That idea tempted me at idle moments in American history class--how far could this go?--but even the defanged patriotic story taught in high school presented enough of the messy events caused by converting borderless, stateless land into invisibly lined and claimed plots, or unlined and simply seized, that the fantasy could no longer give me the same pleasure. It was forcing moral problems on my collection.

On a trip to the Boy Scout Jamboree at Valley Forge, Pennsylvania, in 1957, I acquired Delaware (a view from the bus), Maryland and Virginia (hot, hot, hot), and Washington D.C. (a bonus) which may have been the only reason I had decided, after much vacillating, to go on the trip at all. Seeing Our Nation's Capital (as Bob and Ray liked to call it) and the Washington, Lincoln and Jefferson memorials had little to do with either the abstract Forty-eight in my head or the actual country I lived in. It was just another installment of school, illustrated with architecture. The experience of living for two weeks in a tent obliterated everything else and the most memorable sight on the long trip was endless lines of Boy Scouts snaking from scores of different-colored parked buses to piss in metal troughs set up along Washington Mall behind racks of temporary curtains. I added my stream to the mighty yellow flow and didn't appreciate until I got home that my collection had grown several states richer, besides that bonus of the District of Columbia.

It was natural for a young mind to think the Forty-eight states had always been there since during the whole of my life to that point there had been the solid, stable, yet divisible number of united states. There had been several stages between the original thirteen and my eventual forty-eight, all recorded by the increasing number of stars on the flag, but all had been pointing toward this moment, the present, for fulfillment, as if it was ordained. Even when I learned how the dread issue of slavery got entangled with the admission of new states, those two images remained separate and distinct, a circle of thirteen and a rectangle of forty-eight stars. One marked a beginning, the other an end, an achieved goal.

Then, in 1958, Congress voted to admit Alaska as the 49th state. My first reaction was *This can't be!* The numbers, the design of the flag--it was so right, so perfect. How could anyone disturb those symmetries? Political reasons, or even commercial ones--the power of money--those grubby facts couldn't be allowed to trample an ideal! In my boy's mind there could be no reason big or compelling enough to upset a design of stars and stripes that so properly symbolized the United States. No one would dare. But it soon became official; Alaska was the 49th state. The new flag would have seven rows of seven stars each, an entirely different meaning for numerologists, a base of the prime, lucky number seven with its hint of gambling instead of the stately, stable, bankable twelve times four, six times eight. I dealt with the unthinkable new development by dividing my mind into compartments. In the one holding forty-nine states I discovered that the old lust for more represented by Manifest Destiny had been stored and kept alive. (We might even revive the idea of taking Canada too, which would also make our states, now including Alaska, contiguous again.) In another compartment was the ideal Forty-eight, now designated "the continental United States," no longer a constitutional fact but still an ideal, something separate from facts. So I learned that ideals didn't have to have an actual or concrete form to exist. I wouldn't have been able to describe it that way--I had no language to juggle these thoughts--but could locate the concept of Forty-eight in my mind and by focusing on it produce a certain feeling.

The adjustment to the idea of forty-nine states turned out to be very brief. A political deal had been made and Democratic Hawaii was being admitted to balance Republican Alaska. The new number would not be the unstable product of seven-times-seven but the stolid fifty. As a number it was merely dull, five tens, half of the lordly one hundred, middle-aged, and contained no potent factor like the mystical twelve or the gambler's seven. So in making these changes Congress had stepped lightly over the turbulent stream, putting its foot down for only an instant on the wobbly forty-nine before it jumped to the firm, dry land of fifty, leaving the shining ideal of Forty-eight on the opposite shore where it was still visible but now forever sealed up in History. As well as tucked into my private mental compartment as an archetype of the ideal.

Besides losing the number I'd originally cherished, so confident with its hint of supernatural power, we had also lost geographical unity. Since the states were no longer contiguous, the idea of the union as a solid entity also lost that mental hold on my mind and became even more of an abstraction. The country wasn't defined by a shining sea on each side and the Great Lakes and the Rio Grande top and bottom. Now any place might be a state. How about Puerto Rico, or Guam, or New York City lopped off from the state to become a state of its own? Maybe California should cut itself in half like a worm and regenerate another new head and tail. Why hadn't we made Cuba a state when it asked?

What about the Philippines? The United States could be global. Drop "of America" from our name and call it the United States of the World. Why should tradition or history bind us to anything? We were meant to be new and flexible, that's what it meant to be American. The United States of the World! It sounded like a cartoon. BIFF! BAM! POW!

Outside my mind, out in the sun or the leaden light of an overcast day, this process of increasing confusion over our identity was education itself, in the Henry Adams sense of the word. My mind which was so easily seduced by an ideal was learning to tolerate ambiguity and incompleteness. But "tolerate" sounds as if it was some kind of Enlightenment adventure and I was the willing and happy puppet whose brain was being stretched and molded into an irregular adult shape. Not so; I hated the process. I loathed it, despised it, abominated it. Had I had any power to resist it or still better to change things so instead of having to alter my mind to fit the altered world, the world would conform exactly to the regular polygon of my mind, I'd have used that power instantly, no question about it. What did I care for reality? I wanted my ideal. It was only as a practical matter, to get along, to survive, that I was adapting. My brain, or my mind, was changing in spite of me. Looking back on it now, or looking down on it from my perch I should say, this attraction to consuming an ideal so thoroughly that it becomes a part of one's own self, yielding or even abdicating some of that self to this ideal, seems to be such a part of human nature that it's only in this isolated form that it looks remarkable. I also realized that if I'd been confronted by a powerful agent who tempted me by saying, "Listen, all you have to do is wish for it and I will return Alaska and Hawaii to their territorial status, and restore your cherished perfect rectangle of 6x8 stars on the flag. No? What's wrong? Now you're too fond of the Fifty? You can't give up the two new states! That's very amusing."

26. Intoxication

Jerry was another of those "good kids" hired to work at the soda fountain in John Betts Pharmacy. The primary quality of the ones chosen for this honor--and it was not easy to find a regular part time job that paid a fair wage--was that they, we, weren't the type to hang around on the sidewalk in front of John Betts' drugstore and scare away customers. We didn't slouch in James Dean or Elvis poses against the fenders of cars, or wear our hair in a DA ("duck's ass"), the sides long and slicked back, or in butch flattops standing straight up with wax to create a look that intimidated old ladies and enraged the pharmacist. We didn't exude any sexual message or a surly alienation from the world of adults. Fortunately, however, the soda jerks were not meant to be choir boys either, not nerds or dorks, but more like someone you'd see in a Norman Rockwell Saturday Evening Post cover. Jerry was pretty close to the ideal kid John Betts wanted to see behind his soda fountain. He had a round face with apple cheeks, twinkling eyes, a quick wit and he teased the girls and charmed the older customers, besides being able to talk to men respectfully while still making jokes. To top it off, his father was one of John Betts' best friends, another Past Exalted Ruler of the Elks Club, a man who enjoyed a good time and possessed all the easy social skills reincarnated in his son. There was a final trait Jerry shared with John Betts that made him special. Betts and his wife were childless; Jerry was motherless. His mother had died of an illness when he was a young child and at home it was just him and his humorous, tolerant dad. They lived in an apartment in a complex beside the public golf course, and while their home was always tidy and well kept, it felt strangely uninhabited. Also unusual was that Jerry never asked permission to go out. There was rarely anyone to ask and when he did go out, he locked the door. I didn't even own a key to the house we lived in. The adult freedoms Jerry enjoyed made me envious at times. Then, scrambling this emotion, something happened the night we got drunk together that made me feel there were things I didn't understand about Jerry. Deeper things that made me feel a chill wind blow through his life.

A couple of the regulars at the soda fountain--friends of John Betts and Jerry's father--had been gently nudging Jerry and me toward an initiation into the uses and mysteries of alcohol; they weren't going to intoxicate us themselves, of course, that would have broken an unwritten rule that this step into the adult world should be taken privately, on our own initiative. It would also have been wrong to do it alone since drinking was a social activity and only became the evil its critics called it when indulged in by a soloist. Jerry and I picked a particular Thursday night when Jerry's father would be out for his regular card game and neither of us had to work the next day at the drugstore.

The cupboard in the kitchen Jerry opened only had one bottle in it. The label said Four Roses and it had a bright red ribbon stuck around its neck. "I think Dad won it at some Elks Club raffle," Jerry laughed. He set it on the counter. It had never been opened. "He doesn't keep anything here. He never drinks at home."

"Is he going to mind if you open it?"

"You mean is he going to say, 'What're you doing getting into my booze?" Jerry smirked. "He only drinks Scotch." He addressed the bottle. "What do you think, rye and ginger?" He poured out two drinks over some ice.

There was an outdoor patio off the small dining area with a couple of wrought iron chairs and a table between them. We went out, sat down and lit cigarettes. The drink tasted almost sweet. Medicinal and fizzy from the ginger ale.

"I don't think I've ever sat here before," Jerry chuckled. "Somebody must have given us this furniture. I can't imagine the old man buying it. No, way, José."

It was a perfect early summer evening. Warm but not hot with a basking breeze. The sky was full of stars. I was making sure I didn't drink too fast. That was one thing I'd heard about alcohol, that it snuck up on you. Conversation with Jerry was always full of wisecracks as we tried to top each other showing that the world had no idea how absurdly funny it was. We were soon laughing, seduced by our own wit and merriment. When Jerry said, "Ready for a refill?" I realized my glass was almost empty.

Jerry filled them up again and suggested a stroll on the golf course. I was surprised to feel off balance when I stood up.

"Oh, yeah!" I said to steady myself, as if this was just what I'd expected.

Jerry was enjoying his ability to make himself laugh and I could hear him effervesing up ahead, as if he'd found something incredibly amusing. There was nothing but an old bench near a tee and the dry, scarred ground. The public nine-hole course was barely maintained by the city, used more for sledding in winter than golf. I set my glass down on the end of the bench and went off to pee behind a tree at the edge of the fairway. While unzipping, I looked up, and for the first time in my life saw a star-sprinkled sky. That's what it felt like, as if I had never seen it before--how could I have lived so long without noticing how immense it all was, how staggeringly impressive. The sight of it obliterated all other sensations, insisting that this was the way things really were, if only you looked, if only you looked. How could you not? If you pulled your eyes away you'd miss something unique, irreplaceable. A cosmic event.

I walked back to the bench where Jerry was sitting on the far end, facing toward the apartment building. No one else was out--how could that be on a night as spectacular as this! I found my glass and sat down. It was hard to tell in the dark but it felt as if Jerry must have filled it up again. I was still waiting for something really noticeable to happen, some rupture with ordinary reality when the word *drunk* suddenly occurred to me as if I'd never even heard it before. No, there was no line, no signal. You only noticed the change once it had happened. After it went by. Then you noticed. Although it didn't seem like I'd had enough to drink to be affected. This didn't feel like a mood but the way things were, the way they really were. Then I understood once again, as if it had happened some time ago, that that was the way it made you feel, that the point when it began to have an effect had already passed when you noticed it.

I reached out to get Jerry's attention, to tell him this because it was so funny. It was too dark to see Jerry's face but something must have happened because when he turned and I started to speak, trying not to laugh, I saw what looked like an arm with a fist on the end of it coming towards me. Jerry! What the hell? I was already saying it:

"What the hell?"

The fist missed by a mile but what the hell was Jerry doing? He swung again and I put out my arm to block it. With my other arm I pushed Jerry's shoulder.

"Jerry, hey man, what's wrong? What the hell you doing?"

He kept swinging in a strange slow motion. I stood up to get away from him and he got up too. One fisted arm after another pursued me as I moved back from the bench toward the tee. Then, just as abruptly as it began, Jerry turned around, went back, found his glass which he hadn't lost or broken, sat down and leaned back and drank.

The same slow motion that affected Jerry's fists seemed to be inside my head because I was unable to chase down and articulate to myself the incredibly vivid thought I had about Jerry's weird behavior. It wasn't until the next day when, despite a headache and queasiness, it came back clearly: Jerry tried to hit me because he'd lost his mother--and I hadn't. That was the kind of nutty thing my mother herself might have said, but it had a twisted logic to it. Or illogic. The whole thing was weird. Jerry and I always got along effortlessly, seeking each other out for the pleasure of trading wisecracks on our reliable theme of the silly, self-important, hypocritical world.

When we went up to the drugstore the next morning--as if we had to show off our hangovers to someone who would understand--we were greeted with smirks. Jerry's father came in--had he been waiting?--and when he asked Jerry if he was feeling okay and Jerry said, "Got a stomach ache. Must've been something I ate," Mr. Betts said, "Sure it wasn't something you drank?" setting off a round of laughter among the adults. The predictable comment made it feel as if they'd just been waiting for us.

The two of us, a motherless child and his bewildered friend drunk on the golf course under a spinning starry sky, became an emblem of a particular kind of ignorance. You never knew what emotions were there under wraps, how they might be affecting a person in indirect ways. That was allegedly what psychology proposed to explain. In this case it was my turn to be normal, and normal meant an absence of grievance. By having a mother I didn't know what I wasn't missing. You aspired to be normal so you wouldn't feel anything. But there might have been some other reason Jerry suddenly started throwing roundhouse punches, something way more irrational. It was tiring to try to figure things out when you didn't know enough to come up with an answer. Any answer was a guess. Which was why people made things up. That was frustrating too because you had to fool yourself into thinking you knew something with a certainty that wasn't there. One's mind could make up an answer without your help. It abhorred a vacuum. That was where all the alleged meaning came from. When you barely knew anything.

27. Weights and measures

The metric system was created by the French during their Revolution. The new government recognized a need for a system of weights and measures that would be the same throughout the country, in theory at least. Besides the practical need, there may have been an appetite for order of some kind as the ancient regime was being swept away in a riot of chaos and violence. The meter was originally defined as "one ten-millionth of the distance between the north pole and the equator on the meridian passing through Paris." That last clause is the only part of the definition that refers to anything real, and it's not exactly a singable tune, so it's no surprise that the new system was unpopular. Over time, however, it was recognized that as the commercial, social and--not least important--scientific world became more connected, more "global," a universal standard of measure was needed. It also turned out that the tiny fraction of a meridian passing through Paris which the formula designated as a "meter" was just about the length of half an average human being--male not female--so there was at least an indirect relation to our own dimensions. To give people something to look at which they could call a meter, since even the Marquis de Condorcet's fine mind couldn't imagine one ten-millionth of a meridian, they cast this abstraction in the form of a steel bar which in a later time when the metric system had been adopted universally, was enshrined in a climate controlled sanctuary in Sevres, France, under international supervision. There it sat in pure isolation, protected from anything that might alter or harm it, as if it was a sacred relic instead of a symbol of the new scientific, secular age dawning on mankind.

Beyond the practical aspects of standards of measurement, there is another human need expressed by the steel bar in Sevres, the need for something that endures outside and beyond our mortality. A bar of steel seems like a cold gesture but anything that connects us to something greater than ourselves is worth our respect, if not our reverence. In this case, it's a connection to universality. Any place on earth a meter is a meter. It is a man-made fiction we have imposed on or laid over nature. It is also related conceptually to nature's own phenomena which we can use for standards of measure, like the speed of light. It may one day be exceeded in speed yet the speed of photons, for us today, is a constant reference point, perhaps even an absolute, or as close as we can come to one in our descriptions of nature. What Einstein discovered at the turn of the twentieth century has come to be called the theory of relativity but it has also been called the invariance theory because his insights about light and inertia were so fundamental. There are other physical qualities about existence which manifest these unchanging values, like Planck's constant. These are forces we can't directly experience but their existence is evidence of a world all around us and within us that our consciousness can't comprehend. They exist and endure without us or despite us and besides their physical reality they have the metaphysical qualities we associate with eternity, gods, and religion. That fact should be enough--not to make us fall on our knees--but to open our eyes in wonder, and to contemplate, at least for a moment, our own puniness and transience.

Well, yes, that's a nice sermon, but a portion of the wonder we may feel is a chillier sensation as we absorb into our bones the realization that these physical constants have nothing to do with us as living creatures. They don't know or care anything about us and have no appreciation for us as sentient beings. Or for any other kinds of beings either. If our purpose as living vectors of consciousness is to animate the universe--bring it to life--we seem to be failing our audition. Or maybe not. Maybe there's no other way but this radically irrational process of climbing all over each other, poking our fellows and ourselves in the eye, fouling our own and only home and making it unlivable for so many other creatures, governing ourselves poorly and selfishly, and giving our allegiance to the least worthy among us. How have we come this far? I mean in material progress, in development of technology.

Of course it could all be a prologue to a failed production. Maybe we haven't really come far at all and The End is just up ahead.

28. Refugee

We were seated alphabetically in home room so Ted was right in front of me at the end of the row. He was conspicuous because he was wearing a long cast on one of his legs. "I broke it skiing," he said. He was moving clumsily on crutches (but enjoying the attention) and the teacher asked me if I'd carry his books for him since we had the same classes. "Sure," I said, glad to have something to do on a boring first day of school. It meant we got to each class late because Ted moved slowly on his new crutches and I hung back with him in case he needed help. It also meant I heard a few details about his background while we talked on our slow progress through the crowded, then emptying halls.

Besides the temporary disability, Ted was noticeable because he was foreign. His clothes, his manner, most of all his English accent. His parents had gotten out of Germany just before Hitler closed the doors in 1938, he told me. Ted had been born in London in 1942 where his family landed temporarily after they escaped from what had been their home in Germany. A couple years later when the Germans started launching V-1 and V-2 rockets from France, the shock waves from their explosions knocked him over as a toddler learning to walk. His family, by then including another son, came to the U.S. after the war and his father taught math at a university in Michigan, while his mother worked in the zoology department at Syracuse University. Ted had a survivor's stoic wryness and described his history without any other emotion but there was always a sharp humorous twinkle in his eye, as if he knew life wasn't as serious as people thought it was--particularly here in America--and at the same time you had to be ready for anything because life was also much more serious than Americans could imagine.

Ted was more developed as a person than anyone else in the class and as distinct as he was as an individual he was also the first of a type I would meet in many different incarnations. He was the non-religious, family of man skeptic who, if he believed in anything, believed in science, freedom, and human rights for all races, colors and kinds. It was the kind of person Hitler thought of as a prototypical Jew--the internationalist whose allegiance was to ethical values, not national ones--and whether he was ethnically or religiously Jewish or not, he was the enemy to be exterminated. It was our fate to live in a time when this ultimate immorality had been carried out, painting stark lines around certain subjects. This made someone like Ted a phenomenon, to me at least, although he had absorbed American culture quite easily.

Nevertheless, Ted had few friends among the many other Jewish kids in our school. The difference between them was both social and cultural. Ted's appearance with his dark hair, prominent nose, and hollow cheeks could have let him pose for Shylock, and his lack of interest in, or money for clothes, his wiry physique and minimal athletic skill, made him very different from the stylish fraternity and sorority members in our class, both Jewish and gentile. To them Ted was an embarrassment. He wasn't religious at all, but socially and historically he was the very kind of person the Jewish kids did not want to be. He never expressed any social hostility or disappointment due to his low status; all that was simply irrelevant to him. If he had a group he belonged to at school, it was the art students. He was a good draughtsman and his salty, sardonic opinions suited the atmosphere in Mr. Gordon's art class. The one trait of Ted's that seemed odd was his willingness to fight. It confirmed his fierce independence but also got him into trouble because if someone insulted or challenged him, he confronted the affront head on. If that led to a fight, he didn't back down. He even fought Tony Duma, a football star, who was much bigger than Ted, and just as willing to fight. Naturally Ted got whipped badly. When I asked him about it he made that tilted-head smile, and with a twinkling eye admitted it happened. "Jeez, Ted, are you crazy?" I said and the smile broadened. Everyone knew he lost the fight, but everyone also knew he had done it, that he had the courage to go up against anyone.

His self-possession was bone deep. Despite his English accent, vaguely foreign manner and unembarrassed singularity, Ted never acted as if he didn't belong where he now found himself, in an American middle class high school. In biology class, Mr. Bongo let Ted skin a rat to show us some basics of anatomy. As a natural outsider Ted was oblivious that this was such an uncool thing to do. He was amused when the girls gasped as he slid the whole skin off the dead, still partially frozen rat, and held the parts up for all to see. Behind the bravura show--Ted liked an audience--was a serious purpose. He intended to become a doctor, and he'd learned the dissecting skill to that purpose, as well as with an honest curiosity about anatomy, not just to nauseate classmates. Mr. Bongo, who was Italian, not Cuban or African, and so rounded he actually looked like a drum, squelched the teenaged histrionics at how gross it was to skin a rat, and treated Ted like a colleague who had something to teach us. Because of Mr. Bongo's earnest dignified manner and Ted's serious pleasure in the anatomy lesson, by the end the mood in the room had changed enough that Ted had earned some respect for his ability and particularly for his poise, which was definitely cool, even if he was not. Mr. Bongo's own undisguised earnestness could be amusing but it was impossible not to like him because he clearly cared about teaching and treated us all with respect, as little as we deserved it.

Ted was a reader and he introduced me to George Orwell's writings. I would have learned about *Animal Farm* and *1984* in due course, but not *Down and Out in Paris and London*, which was an introduction to the underside of modern urban life. The cockroach existence Orwell endured as a dishwasher in a Paris hotel was not something I thought I could have survived, but I had no doubt Ted would have. His resilience and sense of purpose seemed all his own, although when I met his mother it was clear there was a source. She lacked his humor which meant her seriousness and brusqueness were unleavened. At the time I didn't stop to think that getting out of Germany and making it to England and then America with her husband and two sons was probably not the first risky ordeal she'd made it through. Now she and her boys were living apart from her husband because Detroit was the only place he'd been able to find employment teaching college math and she had her own job to hold on to. It was an inconvenience taken in stride. Ted said little about his father and I don't think they'd seen much of each other because of the disruptions of war and making their way in a new world. The last year in high school, Ted had no home and lived at the YMCA downtown. His mother had moved to Detroit to join her husband, leaving Ted behind but he seemed quite unbothered by the new living arrangement. In spring of our last year he got accepted by Reed College in Washington state but was going to enlist in the Navy instead. He hoped to train as a medic, go to college on the GI bill when he got out, and then go to medical school. He was quite philosophical about giving up four years of his life to the U.S. Navy if he could swap it for a college diploma. It would have helped me if some of Ted's maturity and purpose had rubbed off but it didn't. His depth of character and sense of himself was enviable but I couldn't simply copy it. The roots went too deep.

29. Emergent

In the line from algae and single celled creatures to hippos and ravens and everyone else, there is clearly an increase in complexity. Isn't that what we call progress? Yes, no doubt. But the scientists have to be careful not to interpret evolution as if it is directed. That's what crosses the line into heresy. (Did I say this before? It's always in mind.) There is no invisible hand--none yet discovered--guiding evolution this way or that. Descent with modification, as Darwin called the flux of evolution, occurs in response to the pressure on species to survive by adapting to conditions in the natural world. So if intelligence helps a primate adapt to conditions in its environment then the trait is favored over time and more intelligent versions of the primate should survive. Neanderthal gives way to Homo sapiens. And now comes another factor, genetic material swapped back and forth at the microbial level, adding a shortcut called horizontal gene transfer to the mix and stupendously multiplying the ways heritable traits can be shared among beings. But the process has no brain itself, no mind of its own, no intention besides surviving. It is us, it's the scientists studying fossils and nature who are trying to figure out what happened to create this world and discover if it means anything. So far, it doesn't. And the scientists can't really say anything about the force driving evolution except that it seems to be universal, the urge of living things to reproduce and continue to live. Plenty of living things exist or go extinct by accident--by mutation, drought, flood, an asteroid collides with the earth--but nothing exists casually, without making an effort by investing energy in survival. What is this great force and where does it come from? If you were looking at earth before organic life bloomed would you see it? Sense it in some way? Or did it only become manifest once something was born--once something sprang forth or oozed out or was sparked into a living form and tried to stay alive; was it only then that the "will to live" became apparent. Was it there before, latent in the energy in the universe, in light and heat which somehow was captured and animated in some combination of nucleic acid, water, and air?

If you look at the big bang and the condensation of stars out of the cooling gas before the eventual creation of an astronomical pipsqueak like earth, there is an enormous gap before life appears, and nothing explains why it did. It may be no more than what scientists call an emergent phenomenon--something emerges unpredictably from conditions present in a certain situation. Unpredictably because you couldn't see it before it emerged. The life force and the power behind evolution may be no more than a way we have of describing what we see, or imagine, looking back at the cosmos from a pyramid in space. Life may have been something no one could have described until it happened, had there been a witness to describe it. There was no rabbit until it was pulled out of the hat. If you could put it back in, it would all disappear--the rabbit, the hat, the hand putting it back, the whole emergent thing, arising from what some of the ancients called the Word.

TWO

What comes into view from the pyramid affects everything visible. Even the invisible. It's also a view incompatible with the very different one seen from the garden. I sometimes feel like the two views are gazing at each other in mutually exclusive bemusement--what are you doing out there, or down there? Standing in the garden, the view is of black (if it's wet) or brown topsoil, and stones--this is New England--and growing green edible plants in season. Surrounding the cultivated plot is a hayfield bordered by a parade of tall trees, oak, birch, maple, pine, spruce. When you are here in the garden you are usually working, doing something that needs to be done, and that need seems to press the garden up in your face. Everything about raising vegetables shortens perspective. It's like taking care of a baby or an animal. Each chore is urgent and necessary. When I'm out here I know what I'm doing, I know what season it is, what time of year, the hour of the day, the chore of the moment, the goal of the effort. I know who I am, am not pleased or displeased with myself as long as I can get the day's work done, and know that there will always be enough left over, yet undone, so that tomorrow, depending on the weather, will present the same earthbound answers and questions as today. This is limiting and demanding but also brings with it a specific satisfaction. It feels essential.

To be at large in the city, out there competing in the technological world, was the way I lived for so long it became reflexive and normal. I could talk like a salesman, make appointments and deals, negotiate, cajole, massage electronic recordings of other human beings in their multifarious activities into television programs. The person pushing a wheeled hoe up and down the rows of corn is someone quite different from that former self. He is self-sufficient, or perhaps I should say self-contained, since he still relies on some help, and all the former desires to make a mark on the common tablet our species etches itself on have become sublimated into something else. I'm not exactly sure what the new state of mind is when the views from the garden and the one from the pyramid *volante* are combined and that may be part of what keeps me returning to the keyboard typing hieroglyphs on a screen. You don't write for no one to read, so who is it for? Maybe I'm just trying to gather some scenes around me the way you pull the blankets up around you in bed.

That the urge renews itself without a clear answer why conjures up the image of Dad at ninety-four, pushing his walker down the hall at the Iroquois Rehab Center, moving almost entirely on will power. Walking beside him down the hall, guiding the walker around other patients and a trolley loaded with medical equipment while he pushed on with such determination, it seemed as if age and illness had flayed him down to the core; all that was left was the raw urge to move, to stay alive. It felt as if he and I, both of us, were witnesses to the primal force driving him, that I probably had it too although mine was disguised by health and relative youth. What I could also see clearly was that if he hadn't lived so long I'd never have seen this aspect of him. I'm glad for the insight, yet he might be content after this walk through the rehab center to go back, get in bed, and close his eyes for the last time. He had no need to be impressed by his own determination. It was in response to--perhaps even a protest against--this form of incarceration.

2. The Lively Arts

Anchel purred and murmured. He didn't need to speak up, God had good ears (even if he rarely bothered to listen), and since everyone else had their own load to carry they didn't want to hear about yours. There was also a good chance the person who might hear you complain was the source of the complaint and if they did overhear it might just make more trouble. So he murmured, he purred. A murmur suited his voice too. Pitched in mid-range, it had soft edges, the opposite of stridency, a lulling sound that made you think of taking a nap. "Take it easy, take it easy," Anchel murmured to no one as he moved around the tables in The Lively Arts coffee house. With his murmur and graceful movements, he was like a cat, a privately wise, not unfriendly tabby. He never moved clumsily and whatever he did was done economically, with minimal effort. If a physical chore could be put off or done by someone else, then put it off, let them do it. Anchel had the kind of natural coolness conferred by having lived a life in which nothing had really worked out. Yet he was a competent man. Not unintelligent. What had happened? Who knew.

His coffee house was a novelty for existing in it's location, so much so that outside the university environs it had to be explained to people what it was. So they wouldn't think something from space had dropped down on Genesee Street next to the laundromat and theater. Anchel himself wouldn't have known what it was if he hadn't created it. His wife had to explain to him what a coffee house was. It wasn't meant for the audience Anchel had to please as head of catering at the Jewish Community Center. If you wanted your event at the JCC--a wedding reception, birthday party, bar mitzvah, whatever--they gave you Anchel's number. Call him. He'll take care of things. Now he was trying to set up a joint to appeal to beatniks. Why not? What did he know.

It was Marie's idea all the way. She picked it up in the Village when she went down to see *Gypsy* and a friend took her to the Cafe Wah. "Anchel could do this in Syracuse," Rachel said, over cappuccino and a shared Napoleon pastry. "It seems like a new one opens every week in the Village. Just buy one of those fancy coffee machines and put up some posters or paint the walls abstract, you could do that." "In Syracuse?" Marie said. "You don't know what it's like. We could be in Nebraska. He'd just go broke again." "You got all those kids at the university. Half of them are from New York, already. There's your customer base. It's a fad, honey--bring him down, we'll show him around!"

The Lively Arts was in an old storefront owned by Anchel's friend Sid, next door to the laundromat Sid also owned. There was a bar, then the theater, at the foot of the long, steep hill the university was perched on. Marie taught art in a public school and she designed the interior. One wall was a mirror, which was already there and made the room look bigger. She covered the others in overlapping posters of Degas' dancers, Renoir's flowers, Django Rheinhardt, Billie Holiday, a *La Bohème* ad, a production of Ionesco's *Rhinoceros*, one for Rossellini's *Bicycle Thief*, a bullfight in Madrid, a French chateau, and street scenes in Paris, Rome, Istanbul. The culture of bohemian Europe papered over and attempted to blot out the industrial smudge of central New York State. A place where people could drink espresso, listen to jazz, folk music, string trios, talk about philosophy, poetry, art, socialism, Russian novels, and wear black turtlenecks, sandals and beards as a sign they didn't belong to Syracuse even if they happened to be here. "The Lively Arts," Marie said, testing the name. "We'll give them a place and they'll find each other," she said hopefully. "Sure, babe," Anchel purred. "Feels like a zoo." He looked around at the small mismatched tables and bentwood chairs Sid had supplied, all enclosed in Marie's decor. "Where are the animals?"

"Well, it's not Nebraska!" Marie scolded him with her popping dark eyes.

The news that this new thing existed had wafted to some of the kids in Mr. Gordon's art class at school. I heard it from my friend Larry, and like Anchel I had to ask what a coffee house was. Larry described it with impatient contempt for my ignorance. "Were you born knowing what a coffee house was?" I needled him. He gave one of his explosive Bronx cheer laughs. He knew he was putting on an act. But I was curious. I had to see this new thing.

The whole front was painted black, the name in white on black in angular, vaguely Asian letters, the mullioned windows of the foyer with candles flickering in the panes. Why hadn't I noticed it driving to church which was just down the street? Church, which I was still forced to attend by parents' edict. Although the university was just up the hill this felt like an anonymous part of the city, where I never took notice of anything. I couldn't just walk in to The Lively Arts alone so I took Mitch. With his red hair and irreverent clowning he was even less cool than I was, but he was adventurous and could talk to anyone. "How long has this place been here?" he asked Marie when she handed us oversized menus covered with her exuberant swooping drawings and lettering.

"Just a few months. We opened last summer." Her flamenco eyes flashed defensively. She didn't want it to sound too new, too nouveau.

Mitch went into a spritz about how unlikely--and welcome--a coffee house was for a town like Syracuse. That's what he called it, a "town," as if he was a man of the world. He was doing his talk-to-adults act, doubly absurd since, despite our curiosity, we were such prize demographic samples of bourgeois American teenagers, and didn't know a fraction of what we didn't know.

"It is something new for this location," Marie said helpfully. "We get a lot of people from the university." There were only a few patrons at the other tables, several couples, two guys playing chess on a portable board in front of the mirrored wall so it looked like there were two sets of players.

But no beatniks, whatever they were, no one who would have looked out of place sitting in a coffee shop instead of a coffee house. Except maybe for the chess board.

The big, floppy menu had an irreverent flair and the drinks and food it described were like an advance look at where the country was going in its tastes, although Mitch and I couldn't know that, and Marie didn't either. Exotic coffees and teas, espresso and cappuccino, pastries and cheeses, all available today in any rust belt urban relic with a Starbucks, but back then strange and exotic. The sole exception was "Anchel's Famous Corned Beef Sandwich with homemade Russian Dressing and Kosher Dill Pickle." It was the most substantial thing on the menu and the one item Anchel was really proud of. The rest, to him, was just a fad, something dreamed up by Marie and her arty friends. But the corned beef was Anchel's own creation. It made the enterprise feel authentic to him, as if it was really his own at its root, whatever else it was pretending to be. The word authentic was the key--if there was one--to the whole culture The Lively Arts sprang from. The question of authenticity was on the mind of virtually all the loosely gathered players in The Lively Arts scene, probably because the culture it was poised against was regarded as so inauthentic, so hypocritical, artificial, plastic and unreal. It's soul was conformist, so it had no soul, merely the husk of one. It worshipped money, it repressed sex and the sensuousness of art, it squelched the freedom of the individual to be himself, to express himself, to be different. Authentic meant free, free to be what you were, who you were.

That was the spirit underlying the decor and inspiring the choice of music Anchel played on the turntable set up in the dining room across from the host's table. Meyer, a musician and radio DJ, another friend of Marie's from the local arts scene, did more than just advise Anchel what records to buy, he showed up one afternoon with a box full of LPs. "Just play anything in here, Anchel. They're all hip." Meyer wouldn't take money. "Give me a corned beef sandwich," he said, endearing himself to Anchel.

Anchel had to pay cash for the brand new chromium espresso machine--a lot of it--but the chinaware came from Sid's endless hoard of used and salvaged restaurant supplies, and he gave Anchel a good price. People liked to help Anchel. In the world of food service and small time entrepreneurs he moved in, everyone routinely screwed each other. Not for large sums; there wasn't enough money in play for that, but shaving a deal here and there, poaching clients or jobs, badmouthing competition was part of the game. Anchel himself played fair. If people cheated, there would be shouting, and when people shouted Anchel poured his philosophy on them like warm oil, "Take it easy, take it easy," he purred. "Don't hurt yourself." Life was serious, but who could take it seriously.

The university students who came to sniff and look around the new night spot brought their own mostly New York and Long Island tastebuds, which at least didn't drive away the customers who resembled them, although Maria was hoping to see more genuine beatniks, a category everyone seemed to have an opinion on. Verdicts were based on who was authentic and who was not. The word was like a test of faith. Anchel would have been glad to acquire a few authentically costumed customers, but he never did his hoping out loud. He'd caught the scent of this need for authenticity and was savvy enough to sense it had to arrive unbidden, unsolicited. Besides, it wasn't wise to let the world know what you wanted.

They took a step into the desired cultural milieu when Meyer suggested he could do some of his radio show from the big table in the main room. It couldn't be a live broadcast, that required an expensive hook up, but he could record interviews with artists, actors, and musicians on tape in a more relaxed, "real" setting than the studio, then broadcast it later. Meyer himself looked the part. Tall and undernourished in his dark turtleneck shirt and goatee, his resonant Atlantis accent sounded steeped in culture, and with him in the room it felt as if there was another authentic artist on the scene. Meyer was there once a week interviewing poets, playwrights, painters, musicians and dancers, their comments punctuated by the hissing espresso machine in the background. "Why didn't you call it 'Les Beaux Arts'"? he asked Anchel, enjoying the payment of his usual fee, a corned beef sandwich and a cappuccino, after recording an interview for his show. "Bozart's?" Anchel purred back phonetically. "Because we're in Syracuse, Meyer," Marie jumped in. "This isn't McDougal Street. You'd have to explain it. One step at a time." To look as if she belonged, Marie always wore a black skirt for waiting tables, and a colorful top that did not hide her cleavage. With her pixie-cut dark hair and flashing flamenco eyes, she looked hip herself, artsy and vaguely foreign. Anchel often wore a flat leather cap and moved around the tables with his peculiar feline grace, nodding and winking, wise in the lack of success his enterprises had bestowed on him.

It was a Wednesday night, there was only one couple in the main room and Mr. Novak alone with his sad European face at the small front table he always occupied, smoking and drinking his usual espresso Romano, looking out through the foyer windows at the street as if waiting for a character from his past to arrive. Anchel was in the small kitchen, and Marie at the adjacent host's table, when she looked up to see the tallest woman she'd ever seen--Marie was short--coming up the several stairs that led from the foyer into the main room. A step behind her was a normal sized man with a full beard halfway down his chest and hair nearly touching his shoulders. The tall woman was dressed all in black, her skirt like a drape brushing the floor, and the man wore jeans and a plaid shirt with a leather vest. With the candles all flickering in the foyer windows below and behind them as they ascended, it was as if this unannounced couple were bringing the long awaited authentic coffee house into the room with them. The Lively Arts was being transformed before Marie's eyes! She almost jumped up to greet them, then instead of offering them their choice of tables among the empty ones, she simply asked them to sit down with her, at the regulars' table by the kitchen.

3. Authenticity

John and Sonya were the real thing. Their arrival banished any ghost of self-conscious aspiration still hovering over The Lively Arts coffee house, and they accepted their role tacitly as if they understood that the way had been prepared for them. With their generous natures, they were glad to accept it, and may have been a little amused. Although they represented an outsider's culture from a middle class point of view, they were also good company and easy to talk to. John was a craftsman who could make almost anything (sandals, a house, beer) and Sonya was an artist who like Marie made a living at it. They had spent the summer in Provincetown, where Sonya did portraits in pastels for tourists and John made leather goods. They were in Syracuse house sitting for a friend on sabbatical from the university. Their subsistence life was improvised from year to year, far outside the acquire and spend cycle of the society around them. Sonya was as gentle as she was big, so big (six feet five, she admitted to shyly) you almost missed how attractive she was . Her long brown hair grew down to her waist, her voice was soft and nurturing, her smile kindness itself. She worked on her art, did weaving and crafts, and John would spend the winter months building up his stock of leather goods to sell next summer in P'town.

The two of them were revelations, seeming to exist in an atmosphere all their own. This must have been partly due to the physical impression they made, perhaps also their self-sufficiency. Exposure to some of the Eastern religions may have created this aura too, and Sonya with her size and gentleness could seem at times like an allegory of wisdom and tolerance come to life. I spent as much time as I could with them, enjoying the unexpected good luck of getting a job washing dishes and doing general clean up for Anchel. I didn't really notice that John and Sonya's personal qualities had nothing idle or licentious or atheistic or nihilistic about them, as the popular impression of a beatnik implied. That image had so quickly been turned into a caricature that the label was rejected as nothing more than a gibe or an epithet by those it was hung on. Most of all it was unoriginal, a cliche, and whatever path John and Sonya were on was their own, and that was the point of the way they had chosen to live. Their outward appearance was inspired by a natural esthetic, something that would in time spread through our whole culture, becoming its own cliche too. Besides doing his various crafts, John read philosophy, both Eastern and Western, psychology, science and history. The book he liked most to talk about was the Beards' history of the United States with its critique of the Founding Fathers as economic exploiters of the young democracy. He and Sonya's alienation was from American culture, not from their fellow man. They rejected the bourgeois style partly because it was dull and conformist--Robert Lindner's *Must We Conform?* was a book I wore out in paperback after John suggested it--but more deeply because our American culture, our entire Western civilization, was rooted in exploitation. I was learning this was inevitable if your nation is top dog, and one of my teachers was Mark Twain. I read *Huckleberry Finn* in school, then two surprises I found on my own were Twain's *Letters from Earth* and his *Autobiography.* Twain's life was a great success, but a flawed one, and he spent a significant part of it struggling with the loss of loved ones (daughter and wife) and the crimes his country committed against persons of color all over the world. The screeds he wrote denouncing brutal imperial treatment of Cubans and Mexicans, the Moros in the Philippines, Congolese in Africa and Asians in China were not humorous or quaint. They were as bitter a condemnation of man's inhumanity to man as was ever written. Written not by a misanthrope but by a man who also called himself, "the human race's oldest friend." My tortured, anguished, ambivalent fellow American had also been a socially conventional celebrity.

It's an enduring paradox that when one's mind is at its most voracious for learning about this world we find ourselves in, it is also, in so many ways, a closed vessel. You don't know anything, yet you already know it all. I was as guilty of this as anyone ever has been. But in John I had a tolerant, patient mentor who believed people had to find their own way and not be led or pushed to ordained conclusions. So sometimes we just sat together drinking coffee, me in an eager state waiting to respond to the slightest word from him, John observing a Delphic silence which might or might not signify anything.

All this has morphed through so many stages and forms during the past half century that it's a surprise to notice from my vantage on the pyramid that our country too has changed enormously--yet it has hardly changed at all. It's not news that the revolution, whatever it was--shiny and made of electrified metal and plastic or the mortal stuff of genes and memes--has been recast as a product one can buy in any of our forty-eight, I mean fifty, states. The style John and Sonya adopted, even the values they lived by then, values that appeared to threaten triumphant mid-century American culture, have been so absorbed by the educated, middle class of today they barely seem noticeable, let alone radical or subversive. No surprise that those values, translated into mainstream society half a century later, have more to do with style and manners than inner substance, soul stuff. At the core we are still materialists and probably always will be. All that has been reinvented is a new, more hip form with sexier packaging. Which also means, to be fair, that our style of living has been enriched and that's no small thing if you're out looking for a good cup of coffee or a pint of beer. Nevertheless, the more deeply authentic, the profound and sublime, is always out of sight, somewhere behind John's Delphic eyes.

4. Hot blue dot

What you can see from a pyramid in space, as long as nothing is gliding by to block the view (tick-tock, tick-tock), is nothing like a movie or play with its carefully arranged scenes intending to elicit an emotion. What you see instead, with proper magnification, is all the random human activity on our hot blue dot. Mostly you see humans because there are so many of us now with only a few wild animals left plus the invisible biosphere, bacteria and all the rest of their microscopic cousins. Whether you can see them or not, everyone is so busy, everything so urgent, and each little wriggling biped you pick up to examine tells you how important he or she is, or has a complaint they want to make to the management, or has an explanation for why you caught them doing whatever it is they were doing that embarrasses them so. These creatures are amusing and you can find yourself picking up one after another out of curiosity, although you won't do it for long before you pick up one who's covered in blood and making an unpleasantly shrill little shrieking noise--a lot of them are making that peculiar irritating sound--and it really isn't long before you realize that as different as some of them look from each other they're all pretty much the same. That's the distortion of such a long view because we know better, we know they aren't the same at all. All those tourists are individuals of some kind too even though they're strolling around the pyramids of Giza taking nearly identical shots with their cell phones to send to their friends back home with an ironic message to let everyone know they aren't *too* impressed by these ancient monuments to vanity and fear of death. Or maybe every one you check out is texting "Awesome!!!!!" on their phone in different languages, each sending the same photo to a different address. The pyramids have become a permanent presence in our imagination. Their impact on people is instantaneous and indelible. This fame might be some satisfaction for the large number of anonymous Egyptians and slaves who built them. Not much, just a mote of satisfaction the size of a grain of all the sand surrounding the pyramids. Some experts say we may have that wrong, that building a pyramid was a good job and the workers were treated well. We don't know, but that argument sounds like slave owners saying the cotton pickers on their plantations were well treated and happy, whether they knew it or not. Besides their function as tombs for pharaohs, the pyramids may have been clocks, annual time keepers, aligned by the sun and each other with a hard to believe accuracy for structures so massive. They are positioned on the site within a fraction of a degree of the north-south, east-west axes, so the shadows they cast are in a nearly exact relation to each other. An impressive achievement for a civilization so ancient. Ancient means we consider it primitive. Maybe it wasn't so primitive.

None of the facts about the Great Pyramid (481 feet high, almost fifty stories) expresses the particular way it stimulates our imagination. Just as a view from space makes the pyramid look very different, the view from a different time, like today instead of 2,500 years ago, finesses the moral problem the construction of it might have posed. Would you prefer there were no pyramids in Giza today if it meant the several thousand lives of slaves and workers it must have cost to build it hadn't been spent? Or would you say we're better off exchanging this singular monument to human ambition and vanity for the lives of a few anonymous human beings? Who would be long dead now anyway. Or, again, is my assumption all wrong and the workers who built it were so well treated and compensated they even competed for the privilege and perks of hauling and carving rocks to build the pharaoh's tomb? As I said, there is some evidence the laborers enjoyed decent lives. So perhaps the pyramids don't stand for regal privilege but the desire for a monument to represent order and permanence, a place for a dead pharaoh to reign over the realm of the dead, while they also show how grandiose and bizarre the human mind can be.

Or you could say what it all stands for is that we can never understand the mystery of the zeitgeist, the way it is always different in another place or time from our own, different in a way we can't imagine, and different from any other place or time, yet similar in that we understand its human character. Both strange and familiar, like the way your home is different from anyone else's. Do we recognize our own time and place through some olfactory sensation the way we register home? This sounds like a stretch, but it could be so by analogy, or more literally and scientifically by some combination of senses, because there is definitely something we recognize as peculiar to our own time and place even if we can't say exactly how we can identify its uniqueness.

5. Bebop

"Jazz is..." Wesley paused for effect before he finished the thought: "...freedom." Then he said: "I'm quoting Thelonious Monk."

He went on:

"Monk doesn't use many words, but just to be sure, after he said it he said: 'Yeah. That's what it is.'"

Wesley wrote cultural and music criticism for the university newspaper and was an occasional guest on the shows Meyer recorded at The Lively Arts. I tried to be there for these sessions and always sat off to the side at the host's table where I could hear and see everyone while staying inconspicuous.

"Jazz, what they called bebop, that was the sound of the truth in the nineteen-forties and fifties," Wesley went on, continuing the primer on modern jazz he was giving Meyer's listeners. "What Charlie Parker was playing, and Dizzy Gillespie, Bud Powell, J. J. Johnson, Kenny Clarke, and the other innovators including Monk, the sound they created changed everything."

Wesley was in his early twenties, had grown up in New York City and had recently graduated from the university. Besides having immersed himself in the music, he also had the authority of having a black parent who was a jazz musician himself. His father had been a piano player in dance bands, he said, and musicians had been in and out of their apartment when Wesley was growing up. When Meyer asked him what instrument he played himself, Wesley said, "I'm not a player, I'm a Marxist." Meyer laughed but he also said that kind of wisecrack could get a critic in trouble. "I'm serious, man." Meyer said. "The blacklist is still out there--I know people who lost jobs. The land of the free and home of the brave ain't so free, is it?"

"That's what Monk's playing, man," Wesley said. "But there's nothing overtly political about it. I mean, anything can be political, but I'm just saying--to get back to the music--the music changed but I don't want to scare people off by saying that either. I want 'em to hear Monk and Bird and all the rest. Hear it as music. Because there's so much there, and there's just so many emotions behind what they're playing. Like the feeling--and this is what the whole modern thing goes back to--jazz was a product of black culture, born with the blues, but by the nineteen-thirties it had become one more thing taken from us, from the black players, starting with our freedom. In that decade when jazz became America's pop music, who was the King of Swing, according to the entertainment press and the publicists? Was it Louis Armstrong the man who showed the world how to swing, if any one person could be said to have done it? No, it was a white guy, okay, a talented white guy, one who integrated his own band, perhaps out of guilt, because he always knew who the real king of swing was and what color he was. Everyone who loved the music and danced to it and sang it owed an unpaid, unpayable debt to the black players and their culture which had created it in the first place. But, again, I have to be careful when I say that. Not everyone wants to hear it. Sometimes I just say it anyway," he laughed.

"You're talking about where the music comes from," Meyer said. "And why it went where it went after the swing era, and became the modern sound."

"Exactly," Wesley said. "Some of the critics say the richness of jazz and the variety of its players was a kind of proof that jazz was not just a folk idiom but something universal, an art form that could be borrowed and expanded and passed through the medium of players who had never been black in America, never been black anywhere. But bebop--bebop was something else. Bebop was going to go down to its black musical roots in a new way, to a place not all the players could go."

"You mean it was a challenge to play it," Meyer suggested. "Both technically and musically challenging."

"Yeah, you could say it had a higher entry level than just playing a melody with a few riffs thrown in. This kind of music required the player to immerse himself in all the notes, to know his way around in the dark. It meant something more than just pleasing the ear or moving your feet in dance. It came out of not being free in a country that praised itself as the land of liberty. The music's goal was to play the truth, the sound of the truth, notes that were not distorted by hypocrisy but expressed a true freedom. To do that they had to move on from it's "pop" traits like singable tunes and danceable swing time."

Wesley never noticed the young white kid in the corner, but I thought I understood what he meant about the modern sound putting off some of the white listeners because I felt his description of what they were playing excluded me too. Why shouldn't it? Even if I was powerless myself, I was part of the white culture I'd grown up in which kept the black players unfree. That was the truth I thought I heard in the music. At other moments I heard another version of what Wesley was describing. I heard the sound of the truth trying to penetrate our bleached out deaf and dumb culture. That was the meaning of the music adapted for a white boy, and it seemed to spread out in all directions. Its virtues were unimpeachable: It was authentic, it shunned hypocrisy, it was out there for anyone to hear yet it insisted on its own integrity. It was the sound of something I could aspire to but not imitate; I could never play it myself.

6. Anchel's son

It's exhausting to be young, even to remember it is exhausting. Everything is so true or so false and has to be praised or denounced with such vehemence. No one wants an accurate portrait of those overheated days and nights, and from out here on my perch it's easy to miss things I'd prefer not to see, or even detour into fiction. No, nothing made up (despite the temptation). I'm stuck in reality--if that's what this is--and you know it is the moment your body cries out in pain or hunger or feels sick, pushing everything else out of mind. Right now I feel fine although the wind is blowing violently. This is another new feature of what we've done to the planet's climate. The wind, when it blows, blows harder. The rain falls harder too, almost as if it is angry at the earth and wants to pound it flat. This is no illusion. Statistics confirm that a larger volume of water is dropping on us. We haven't had a week without significant rainfall since the fourth week in July. The garden was sodden, now the compost heap area is mud. This is the new, more extreme New England weather. So I'm glad to go back to spy on the young couple sitting at The Lively Arts some time in the twentieth century (but only for a glimpse). The young lady is having jasmine tea, the young man a double espresso. They are going to share the lemon tart Anchel is putting down on the table; as he turns away he winks and purrs, "Pretty girl, pretty girl."

Anchel is charming and Karin is amused by his flirting, so why is she uncomfortable here?

It's an alien place. She can feel it trying to push her away, to make her feel she doesn't belong here. The beats are dirty; they don't wash. That's just stories in the newspaper, he, I, her escort tells her. Trying to shock people. There aren't any card-carrying beatniks here tonight anyway (like most nights). Even John and Sonya don't qualify except for her long, long hair and draping black skirt, his full beard and leather vest and--that's quite a few exceptions. But they are clean. And they aren't beatniks. Beatnik is one of those human types no one actually is because it was coined as an insult, to mock certain people and for them to admit to being such a thing would be to acknowledge an identity assigned by the straight, uncomprehending, philistine world. In other words, to be inauthentic.

Karin's family is more tolerant than average perhaps because they originally came from Holland and her father is a believer in civil rights, world peace, and socialism, but she is also quite thoroughly an American girl. For her this place is something different. Something strange and discomforting. She said the same thing about it as Mom did, "Those people are dirty," although not with the same quivering hostility. Mom meant they were not just unwashed but immoral. Not our kind. So soon the spell will break. They will forbid me to work here. I will lose what is both a refuge and an identity, and cry the hot tears of self-pity and self-righteousness. O lost!

Everything is either too true or too false.

The American Empire was what loomed over us, John said as we sat at the host's table drinking coffee, and the Gross National Product was the output of our daily lives, goods and services, tangible and intangible, beliefs, myths, and prejudices, bombs, tanks, and warheads. In the aftermath of the Depression and World War Two, the industrial engine had taken off, and suddenly we were engulfed in the Age of Affluence. Who could be against increasing the wealth? The discerning individual who realized that he himself was the product, a pawn of planned obsolescence, human fodder processed into a commodity, along with everything else pouring out of the marketplace cornucopia. The promise of America and its democratic dream had been corrupted and was poisoning the population with its artificial substitutes for real food, real experience, real life. John drew his picture in a thoughtful voice, as if he understood and even forgave the benighted populace for not seeing the darkness around them. To see things for what they were required something like a conversion. It was like learning to see in color. While John talked, the action went on around him--other conversations, bebop on the record player, the espresso machine hissing like a dragon, Marie ringing up a check on the old oversized brass cash register. On the nights when he was in the right mood, John talked. He was opinionated but temperate and aware of my naiveté and ignorance. He respected the former and tentatively offered his own knowledge as an alternative to the latter. "Some things he doesn't talk about," McBean told me. He knew John from years past and they were contemporaries. "Ask him about Reich and his orgone box," McBean laughed. "Crazy." John didn't want to be the one who disillusioned me about my country and its imperial crimes, the hypocrisy of society about money and the predations of free enterprise. Racism was different. We all knew about that, our indelible stain on our history, even if only superficially in my case. John might have been protecting some residual core of his own faith too by tempering his opinions. Whatever his expressed thoughts, he preserved faith in something. He was probably also aware of Sonya's and Marie's frequent presence at the table with us. Sonya could speak casutically about social injustice but her voice was so gentle that coming from a person her size it made her seem like a guardian angel, one reluctant to use her great power lest it inadvertently cause harm. Instead she urged us all to do better, to be better. Marie was protective too, although they all seemed a generation older, at least, and when the conversation revolved around them, as it usually did, the questions raised by Jung, or abstract art or George Washington's real estate manipulations isolated John and me in a side conversation. Among the complete coffee house cast, there was also Neil, a philosophy student at the university who helped Anchel in the kitchen on weekends. He made wry comments in his Mike Nichols voice and was too intellectually lofty to talk seriously to a kid. He rarely even greeted me.

The Lively Arts was a closed world. I never saw anyone from there outside the place, no one except McBean who became a kind of friend. With his goatee, long yellow hair and sandals he was obviously an outsider of some kind and I was glad to be seen walking around the neighborhood with him, up to the university and the bookstores, mingling in the student street life. McBean had been in the military, discharged when he and the Navy acknowledged their mutual incompatibility, one that seemed to extend to any other prescribed role he might try to fit himself into. Unlike John, who was firmly self-directed, McBean was improvising his way from day to day, content for the moment to work at a print shop and be guided by serendipity.

The other exception to seeing people outside the coffee house happened one day when Anchel told me to finish cleaning up later and come with him. I thought he wanted help with something. He didn't say he was buying lunch. Anchel spoke so little to anyone but himself it was strange to be in his company walking up to the luncheonette at the end of the block beyond the theater. True to form, he didn't say a word or explain the invitation on the short walk there and inside he purred his way toward the back where we sat down at a table that was the counterpart of the host's table at his coffee house. A middle aged woman was sitting there with a stack of counter checks she was totaling. "C'mon, Anchel, I'm waiting for you," she called out to him. We joined her, another guy sat down--the lunch rush was winding down--and we ordered sandwiches. They all ignored me and talked business, and it wasn't till we were nearly finished that she said, "So, Anchel, you gonna introduce me--is this your boss, or what?"

"Yeah, he's the boss," Anchel purred almost inaudibly. "No, he's my son."

"Oh, yeah?" she said in the same jokey voice. "Your son?"

"This is Eli," Anchel said. "Haven't I introduced you?"

"No." She looked at me, studied my face, almost taking him seriously. "You're putting me on, aren't you?"

Anchel didn't answer. He just smiled and got up from the table, as if ready to go.

"Are you really his son?" she said to me.

I smiled too.

She shook her head. "That Anchel, he's full of surprises."

The guy who ran the luncheonette had gone back behind the cash register and Anchel was paying him.

"I don't believe him," she said to me as I got up. It sounded as if she was apologizing. "He never said anything about a son before."

I didn't know what to say. Had he brought me here just for this? I was flattered but just as surprised by Anchel as she was. Taking my cue from him, I smiled again and said nothing.

7. The hour before dawn

The image appeared, two hands, left and right, hanging loosely by someone's side, now folding into each other like mates--but the image has gone with all the greater context that flowed from it, leaving only a residue of something gone. At their touch, the hands recognized each other and the scene belonged to them. It was a scene I was filming, and what is odd about all this is that it conjured up a rush of unrelated images on the theme of things I'd done wrong, the movie that so often plays before dawn, especially during the mental crunch of the autumnal equinox (or the vernal equinox too). As vivid as these spontaneous images can be, they flee with the light as if they belong to the darkness. This isn't because my derelictions were so heinous--mostly I hurt myself--and after these glimpses some saving filter hides them away where I can't recall them. Or if I do they don't have the awful potency they have in that hour just before dawn. Instead they seem like something I should have grown out of or rejected from a survival instinct. This may be the protection one earns from years lived, the healing of time, our motion through space, and the changed perspective of the passage through this dimension. They might also be artifacts of the view from a pyramid. Not one posing for tourists on the sandy rock but the one under the spell, the laws of another physics. I could also say that as time passes through us, through our minds and physical beings, as we pass through it, we feel life is less serious than before all those years made their circuit through our being. Yet life is serious, as Anchel said, once or twice in my hearing--not as if he believed it but possibly to remind himself to pay attention, to respect the cliches everyone else lived by so he didn't make trouble for himself. At the same time it's a true fact, as my friend Zeke used to say, that life, having no alternative, is more serious than it should be, until you remember that whenever we look over the edge we are staring into our own absence. And that so far earthlings are the only creatures we know of who can have this experience or express it, unlikely as that seems.

Here I am again, by accident, an accident that may have a dash of inevitability since it seems unable not to repeat itself. I began by trying to recapture a pre-dawn perception that seemed so important just a few minutes ago and now I've forgotten what it was. Oh yes, hands. When they touched they realized who they were. A tactile sensation. Why did it accompany a flood of scenes, coming like flash cards, instances of when I'd done wrong, embarrassed myself, faltered and fell out of my own approval? But that is my character, my idiosyncrasy. I only wish to believe otherwise.

8. Whites only

By adolescence--and I should say here that the collage is not really organized chronologically, or if it is in certain instances it's only an incomplete outer ring. My intention is not to deceive anyone, including myself, and the best way to avoid that is to shun any self-confident coherence. I also want to make sense but not force things into place artificially, so the tracks any structure proceeds along will obey their own rules, even if they aren't apparent.

So, let's start that thought again: By adolescence, collecting states had begun to seem like a childish goal unless it was connected to some larger function or purpose, like doing something in each state. A trip to San Francisco with Fox when I should have been in some institution of higher education added Utah, Nevada, and California. It was my first sight of the Pacific. Like the Spanish explorer Balboa, who supposedly stood on the shore in Panama and named the great ocean which had by chance lain down that day as calm as a pond, we stood as far west as we could go with the Golden Gate Bridge behind us and discovered for ourselves the mighty shining sea. We pretended to be blasé, unimpressed by being impressed. "Big fucking ocean, Fox," Fox chuckled (we called each other by the same nickname, as if we shared the identity) but our coolness was fake. Had we been turned inside out we would have revealed the wide-eyed, overwhelmed innocents within, and we had to mock it for mocking our puniness. But only briefly. Then we were silent; the sight quickly made us mute.

Our destination on the cross-country return trip was Florida where Fox--Brady--went to art school, and we argued about which route to take. I wanted to swing south through Arizona and New Mexico which I hadn't yet seen, or collected, but it was Fox's car, his old powder blue Caddy, and he insisted on backtracking through Boulder to see his girlfriend again before we angled south to Florida. Our argument would have been calmer had there not been an audience but we'd picked up Mose in San Francisco, and although he was indifferent about the route, we were in some degree performing for him. Fox said my route was longer, I showed him the map, it was the same distance. He said there was nothing he wanted to see in New Mexico or Arizona. Wasn't he curious? Had he ever been to those states? No, and he didn't want to go now. He wanted to go back and see Judy. The argument was just exercise. He owned the car. He wanted to squeeze Judy again. What could I offer against that attraction? The southwest would have to wait. We drove back over the Sierra in the old Cadillac, back across Nevada and Utah to Colorado. When we finally left Boulder again at two or so in the morning, Fox and Mose were exhausted, both full of 3.2 beer, and I drove us the rest of the night. As if aware of a purpose, I had drunk only a couple bottles of the weak beer and took the wheel when we finally resumed the long drive east. The wee hour solitude was invigorating; when we sliced through the long, thin western handle of Oklahoma, I was going to acquire a new state (one I'd have missed had we taken my southwestern route) and at sunrise I looked out the car window on the left straight into the massive red solar ball lolling on the horizon and felt the odd sense of purpose confirmed. It made me superior in some undefined way to Fox and Mose, dead asleep while I drove. What purpose? I didn't analyze it. The sensation was enough in itself. The mood soon dissipated; a breath of the moment.

There was no conscious connection between this fleeting sense of purpose and the collection of states. I no longer believed in the original quest to collect, nor in the whole, regular shape it had once had in my head. The forty-eight sided polygon had been a child's toy, a fantastic artifact. The appetite it had yearned to satisfy still existed in my irregular mind along with its ideal, in some evolved form, and the tension between them had effects I rarely recognized. Not for what they were, anyway. Nor was the impulse itself something I could explain or understand any more than when I first noticed it. It was as mysterious as ever. So the gathering of states continued, surviving on the capital of the original ambition and tucked away in a reduced corner of consciousness, one that seemed to get smaller as the collection moved closer to completion.

Several years after Fox and I took our cross-country trip, on another trip west with a woman who was briefly my mate, I added North Dakota, Montana (at last!), Idaho, Washington and Oregon, to make forty-six. Only Arizona and New Mexico were left unvisited (unless you included Alaska and Hawaii, which still felt like changing the rules in the middle of the game). I finally gathered up the two southwestern strays on a trip filming mad Captain Bill as he towed his boat three thousand miles across the face of the landlubbers, from Wall St. in Manhattan to the waterfront in San Francisco (from where he would, improbably, embark for Tonga). That was it: forty-eight. By then the collection and its ruling ideal was a virtual joke, something I'd have been embarrassed to admit to anyone. It had collided with reality a long time before, on the day when I got out of my 1952 Plymouth in front of a restaurant in Georgia and walked in through a door marked "Colored Entrance." I hadn't noticed the sign. When no one came to wait on me after five or ten minutes, I went through another inside doorway to the lunch counter and asked if I could get something to eat. "We'd be glad to serve you in here," the waitress said. Her voice was not unfriendly but firm, unyielding in fact, and when I looked back I saw the "Colored Only" sign over the doorway I had just come through. This state, Georgia, which I'd only recently entered for the first time, did not feel as if it was mine to collect. It was a place I had just learned I did not want to step into, not unless I could do it by passing through the Colored Entrance, either by mistake or intention, and be served. It's too simple to say yet it's almost true that on one side of the line was the boy's world where you could be a collector of ideal states--totems like the neckerchief slides we swapped at the Boy Scout Jamboree with kids from around the country--and on the other side of that line was our very real loving and hating country where the lines were more than merely visible or invisible, but drawn in blood, more to be shed very soon by the Freedom Riders, who would not be sitting down by accident at a Whites only table, but doing so intentionally, to change the flawed world we lived in. By the time I drove across Arizona those several years later and filled in the forty-eighth piece in my personal jigsaw puzzle, I barely noticed I'd completed the old childish goal. That didn't happen until I drove out of the state and saw the sign bidding me goodbye.

9. Apprenticed

The tendency to create an ideal, to host one and let it grow in my mind like a coral, has caused me no end of discomfort and sometimes trouble; apply it to sex and romance, work and purpose, faith and belief, if you want to grow up it has to be outgrown, even uprooted, or simply ignored. It leads the mind into a habit of double vision so that most anything real can have an idealized counterpart, a Platonic Form, and your actual behavior, your decisions and thoughts and mental arrangements, are all dependent on whatever compromise you can make at any moment between the real and the ideal. Even to try to describe it is paralyzing. It feels at once too reductionist with everything falling too neatly into labeled categories, and at the same time it seems too complex because there are so many qualifications and nuances and footnotes to add to any plain statement that a clear insight is soon obscured, as the long history of the conflict shows. And speaking of beautifully wrenching autumn days--which I wasn't--and the several dichotomies of the real and ideal, there was the one that might be called knowledge, or more humbly, education. Autumn reliably brought the anticipation that this year would be different, the unblemished first days, the hopeful gathering of the tools--new books and fresh notebooks, a favorite pen--and then the swift defacement of the ideal school year by the dull classes, the messily inscribed notebook with its stupid doodling, the asshole who wrested a chair from me in the cafeteria, the whole institutional daily rote and routine was mediocrity itself. But here's something that isn't routine (just wait), it's a "Dairy Queen"! A cow with a crown! Producing royal ice cream! It sounds like some kind of ideal itself. It was--to the owner, a Mr. Kay, whose whole life yearned toward a perfection of neatness and cleanliness, and economy and punctuality in a world of dirt and disorder and waste and sloth. The temple to his gods was the Dairy Queen building on its patch of macadam on a wide, lazy street in downtown Sarasota. Among the young kid's other chores--the young kid was, quite improbably, the same one who should have been somewhere in a classroom learning about Western Civilization--but instead was hosing down the front patio in the fragrant tropical air. To senses trained in the industrial grit of central New York, the spicy arboreal droppings that fell from a pair of exotic small trees were pleasant, almost intoxicating, and didn't need to be hosed off every morning, but to Mr. Kay it was essential to fulfilling his ideal Dairy Queen that the patio must be without blemish. Inside, which was even more under his control, every act and procedure was as exact as tea ceremony. For the first two weeks of indoctrination, his young employee was not allowed to touch anything except a hose, a broom and a faucet. All the hardware, materials and foodstuff specific to making soft ice cream, cones, milkshakes, and sundaes was handled only by Mr. Kay, and he described each action each time he did it, with a never lessening urgency, as if Eli (I was not really myself but some third person in this role) were training to be a jet pilot or surgeon. You couldn't transfuse blood into a patient with more care than Mr. Kay took when he filled the ice cream machine from a cardboard tub of what he called "the mix." The central act of all the minutely studied actions was making a cone of soft ice cream. Eli was instructed to watch carefully the technique with which the trained adept opened the valve with his left hand, crossing over and above the wrist of his right which held the empty cone under the spigot from which oozed an even, voluptuous flow of soft ice cream. Of course there was something anal about it; Mr. Kay's personal and professional hygiene were a triumph of Jewish-American toilet training. He assumed, or suspected, that Eli was Jewish too, and made a wry comment about their shared ethos, though they never spoke about it further. Eli's silence was assent. Mr. Kay had a curious mixture of fastidious discretion and bluntness. He wouldn't hesitate to grab Eli's hand and correct it's position, barking, "No, do it like I showed you!" but after the initial interview (Eli "beat out" twenty-five other applicants he was reminded more than once), Mr. Kay never asked another personal question. That was the key, personal. As much as he imposed himself and his comical discipline on Eli as he taught him how to run the ideal Dairy Queen, he was otherwise a private man who respected another's privacy. He had one bad habit too, smoking, which shined a softening light on the lines of his tanned, jagged face. The ashes, fire, and smoke of cigarettes in the back room of the ice cream temple might have made him frown, but instead he seemed pleased when Eli said, yes, he smoked too.

The course of instruction lasted six weeks. For two weeks Eli watched. For the next two, he was allowed to do some of the less exacting procedures, like filling the stainless steel tank of the ice cream maker. At the start of the final trimester, Eli was allowed for the first time to make his own ice cream cone. To try. Failure was presumed, and fulfilled. The flow came too fast, the cone was a tall sagging spiral of wasted ice cream. Mr. Kay's big hand grabbed the disaster and wiped it off on the edge of the tray. "You're not watching what I do!" he said, trying mightily to sound patient. Someone must have told him--his wife, a former employee--that he was overbearing or too impatient or just being a jerk, because Eli could feel the man reining himself in. Eli had stood there watching what Mr. Kay did for a month and for approximately twenty-six or twenty-seven of those days his brain had been either shut off or elsewhere. You couldn't watch a man make an ice cream cone for nearly four weeks and keep paying attention unless you were too dim ever to grasp the blessed technique anyway. Fox thought the whole thing was hilarious. "Touch the machine yet?" he asked whenever Eli came home from the Dairy Queen to the moldy cottage on the bayou they shared. Puffs of exhaled smoke punctuated the air as Fox made the low, vulpine chuckle that was his usual comment on anything touched by the sweet fairy of stupidity. Eli felt frustrated and insulted by this long tuition in the art of making a Dairy Queen ice cream cone. Obviously it sprang from some neurotic need of Mr. Kay's for total control but how could a man who wasn't an idiot make something so simple so complicated? Couldn't you just watch someone once or twice, then try it yourself, screw up a few times, and five minutes later you'd be twirling spirals as fast as the chilled mix could flow out of the spigot? You could. But would you be making the ideal cone? Would it be just the right size. Too much and you cut into the profit. Too little and you sent trade to the competition down the street. It also had to be straight, not a leaning tower of ice cream, and it had to have a neat curl at the top like a cowlick. As you learned the finely calibrated differences in height between small, medium and large, you were also supposed to be appreciating the essential fact that it wasn't as easy as it looked, Fox, especially if you were trying to emulate the Master of Dairy Queen cone-making himself. "At least he let you touch the machine," Fox chuckled. "Even if you fucked it up."

Like everyone who has an ideal etched in his mind but is mired in smirched reality, Mr. Kay lived in a constant state of tension. If anything was ever good enough for him, it was only momentary. His ideal forced him to be vigilant all the time and eternal vigilance made him irritable. But he could also be placated by seeing that someone--his employee, his pupil, his apprentice--was observing and replicating his ritual. Worshipping his god, if you want to call it that. Or catching his disease. Once he saw that he had been taken seriously he could relax to some degree and enjoy a certain level of humanity, even comedy, with his employee. When the temple was clean and there were no customers, they stood in back smoking together and Mr. Kay told him stories about his days as a car salesman. It didn't seem like the right job for the Master of the Dairy Queen, but once he started telling a story and slipped into his jocular, deal-making language, he was just as convincing in that role as he was as a neurotic autocrat. He told Eli he had owned a Chrysler dealership which didn't sound quite convincing--what was he doing running an ice cream stand in his seventies if he'd owned a business like that?--but he also confided that retirement had driven him nuts so he bought a small business to keep himself busy. Eli didn't care how much of what Mr. Kay said was true. It was all plausible, mildly entertaining, and it meant he wasn't repeating for the umpteenth time how to mix the syrups for the toppings while standing over him--Mr. Kay had a tall, vigorous physique--to make sure if a drop of sticky liquid landed on the counter or floor, Eli wiped it up. Every horizontal space had to be wiped down with Chlorox each day when he opened up in the morning, and one morning just after they opened, a woman came up to the window to order a Blizzard, which was what the Dairy Queen corporate headquarters called a milkshake. It was made with so much ice cream it had the viscosity of cooling lava and as Mr. Kay reached over to set the Blizzard on the small semi-circular counter that jutted out from the window toward the customer, the bottom edge of the tall cup hit the rim of the counter on his side of the window. The cup tilted down, disgorging a thick pancake of Blizzard on the freshly Chloroxed counter. Mr. Kay glanced up quickly at the customer who was looking off toward the boulevard traffic and with two swipes of his large, capable hand he smoothly squeegeed the spilled Blizzard off the counter and back into the cup. A second glance at the distracted customer told him he had another second or two which was enough to wipe his hand and the counter with a clean towel and put the reconstituted Blizzard through the small window on the counter in front of her and say genially, "There you are, ma'am!" Eli admired the performance hugely, though he doubted it was a move he could repeat himself, not without practicing. In fact, his admiration was so great that he had to go into the back room so Mr. Kay wouldn't see the way his torso was wracked with convulsions of suppressed laughter.

At the end of the tutelage, Mr. Kay left him in charge of the temple. Every other day for thirteen hours each, Eli was the sole high priest in attendance. He was thoroughly schooled, absurdly knowledgeable, but Mr. Kay was the kind of teacher who instead of making his pupil feel competent made him anxious he wouldn't measure up, and since that was Eli's natural composition--to feel that whatever he did he could have done more, or better, or finer, always in the shadow of the unscalable pinnacle of perfection--it was two against none and the part that was left to defend himself was the rebel, the misfit, the malcontent. It was a struggle to keep him under control, but he did, and he did his job, and when he closed up at night and emptied the cash drawer and zipped the money up in the small canvas pouch it struck him that Mr. Kay was more worried about the way he treated the temple than his handling the money. His instructions on that subject had been brief, perfunctory. Following another set of directions, he drove his aging Plymouth through a series a turns into suburban Sarasota, found the ideally tidy house set in the immaculately trimmed yard. When he pressed the bell Mr. Kay appeared so fast he must have been behind the door waiting. It felt like the first time Eli had ever seen him. Tall and tanned, standing in the crevice created by the half-opened door and with a TV murmuring in the background, he held out a hand for the pouch and said, "Everything go all right?" "Yes, fine." "Good. I'll see you day after tomorrow." Not only did he not invite Eli into the sanctuary of his home, he had only opened the door to a width less than he could fill with his body which made Eli feel that instead of delivering something welcome he was begrudged this brief audience because the pouch contained contraband or Eli was wreathed in a cloud of germs or that the delivery had simply interrupted Mr. Kay's ideal privacy, and maybe his favorite TV show.

Eli got back in his car and drove to the moldy cottage on the bayou.

10. Tony

Early one autumn morning, Eli was standing as if paralyzed or struck by lightning in one of the bays of the Foreign Motors garage, looking out at nothing, a palm tree, the horizon, a pyramid in space, while holding a wrench in one hand and a greasy rag in the other. At this moment his blanked mind contains no coherent thoughts. It resembles a petit mal epileptic seizure when all consciousness seems to be arrested, clocks stop, and pyramids pause in orbit.

Eli is dressed in industrial whites with a name stitched in red thread above the shirt pocket. The name is "Tony" which is not Eli's name, nor is he any longer Mr. Kay's apprentice. After six weeks of tedious, hilarious (if you were Fox) training, Eli had to tell Mr. Kay he was going to quit. It was one of the hardest things he had ever had to tell anyone. Why did he care about Mr. Kay's feelings? Who was Mr. Kay to him? Who was Eli to Mr. Kay? Was Mr. Kay standing in for Eli's father? Like so many questions, all these do is falsely imply there was an answer. At this point in his distracted peregrinations, if a chance acquaintance offered Eli a job as an apprentice mechanic, which is what had happened, Eli had to take it because he had spent a significant slice of his life fascinated by cars and internal combustion engines. Against this unsought opportunity to get paid for working on them, and taught their mysteries, an obligation not to disappoint Mr. Kay collapsed. It might be interesting to digress into the psychology of Eli's desire not to let Mr. Kay down--not to let anyone down--but it would be an unhealthy pursuit. Mr. Kay was indeed crushed, briefly, his suntanned face crumpling as the full weight of his acolyte's defection fell on him. Eli crumpled too, though less visibly, as his Dairy Queen mentor joined the list of those he'd recently disappointed on his downward progress.

Speedy, owner of Foreign Motors, breaks his new employee's stoned spell by shouting, "Hey, man! What're you doing?" Eli jumps to attention, recalls himself from a distance, and continues across the shop. In a newly learned maneuver, he drops down and lands on his shoulder and hip on the low dolly, his momentum causing it to slide under the car he's servicing, a Triumph TR3 perched up on jack stands.

11. Eight and a half

Fox and Larry both rejected success. They did it pre-emptively, without waiting to see if it happened, or doing anything to court it. Fox perhaps because he saw what his father's fame as a research scientist meant to the quality of his life (little or nothing), a confirmation that his father was right to follow his instinct for modesty, a product of his shyness, and a desire to live closer to the truth about man's place in the cosmos (although in the character of that modesty his father would have made a joke about it instead of advertising it). For Fox himself, who wasn't shy like his father, scorning success was also partly a pose, part of his coolness. Nothing could impress him that was meant to impress him. He was the arbiter of what was impressive. He was glad to hear praise for his painting but it didn't affect his own opinion of his work. (Or if it did, no one knew.) In Larry's case, his precocious gift for drawing in pen and ink seemed to lift him up and over everyone else, including most of the other art students, as if he was already an adult, or had an adult skill and not a common one but a rare one too, one that was unusually developed and refined. This skill wasn't the result of hard work and dedicated application but had been bestowed on him like a blessing, almost fully formed, by the touch of some magic wand, and nothing he could do in his life would exceed it. That was the Faustian exchange: it was an unmistakeable honor to own this ability to create images of life--not only an artistic distinction but a spiritual and even a religious one, to be able to conjure something from nothing in imitation of the original act of creation. But having received the blessed bestowal, what was left for him to do? To aspire to. Nothing, except to nurture this gift, curate it, care for it and use it appropriately. Of course Larry didn't sufficiently respect his unearned blessing. How could he when the only way he could claim his own individuality was to refuse the full honor of the extravagant but finally limiting gift. Besides his artistic skill and temperament, Larry was also a big guy with a manly physique (though not athletic), a heavy beard, and very good looking. He always had girls around him, attractive ones, and barely had to ask for a date to get a companion. He'd grown up in a house with women (his father died suddenly when Larry was a boy) so he knew female culture and could speak its vernacular, talking to girls in their own gossipy chatter, assuming by turns the role of the cute wee boy, the male coquette, as well as the helpless and needy, rogish esthete.

Fox and Larry were more than friends, I also admired and envied them because they had such facile, manifest talent. What I had was desire, but didn't know how to express it since I lacked a natural art. It was another visual art that seemed to answer this yearning in a stunning, unexpected way. On a hot October Florida day, Larry and I walked out of the movie theater in Sarasota where we had just seen Fellini's *Eight and a Half*. We were speechless; we didn't know how to say what we were feeling. Larry tried first: "I don't understand it but I think it was great." Then he laughed that sputtering mock contemptuous laugh that he was always showering himself with. I felt I understood what we'd just seen yet couldn't say why. I didn't have the language for it. It was the first time either of us had seen a self-portrait of an artist which heaved overboard conventions of plot or narrative logic or character development in favor of scenes hatched straight from the raw material of experience, or fantasized versions of it, or the unconscious. Instead of the usual "necessary" scenes constituting a movie so the audience could follow a story about certain fictional characters along a thread of narrative logic, this was a new form that chucked all the dross, all the artificial connective tissue that distracted from the immediacy of experience as it impressed itself on our senses. The possible meanings of a film made according to traditional conventions were unprofound and unresonant because it was comprised of made up material, and that material was limited to a repertoire of familiar cliches, strung along the devices of mechanically clanking plots. Since art school students were all familiar with the modernism of the twentieth century including contemporary abstract expressionists, it didn't take long for the connections to click so everyone could understand what Fellini was doing. What we thought we had seen but lacked words to describe was confirmed by critics who told us this was indeed an exciting new world, and yes, we really had seen what we thought we had seen, and more.

12. Montage

We'd grown up with movies, they were another innovation of our century, and it was strange to discover that the familiar form of entertainment we'd seen Saturday afternoons at the Palace Theater, eating popcorn and candy while watching newsreels, cartoons, and Westerns, had suddenly become an art form. Not quite so suddenly as it might have seemed to us in our ignorance, but swiftly enough to be a kind of revelation. Part of the revelation was learning how far ahead of us the Europeans were. Movies for them weren't just popular entertainment, they were taken seriously, and even had a vocabulary to describe esthetic effects, terms we hadn't heard of.

The most distinctive of these, perhaps even unique to *le cinema*, was what the French called *montage*. It named the junction between two pieces of film, the invisible line where two scenes in a sequence were joined. When one scene was replaced by another, the viewer might not notice anything more remarkable than a change in camera angle, a transition from a wide angle shot to a closer one. Or more likely not notice anything at all when film technique was used to create apparently seamless storytelling. However, in other instances, the intention was not to create an unnoticed transition but to seduce or jar the viewer's attention, even to shock the eye with something unexpected and dramatic. In these more extreme cases of montage the viewer could be tricked into feeling an emotion or esthetic response to something that was not even there. Montage did this by editing shots together so what you expected to see appeared in your brain though it didn't exist on the film. In other words, the movie was inside your head in such a way that you, the viewer, were even part of the creative act. This phenomenon was set off by the jump from one image to the next. In this invisible space, "on the cut" in the film editor's parlance, were all kinds of potential emotions and thoughts, unique to each viewer yet also shared--the ones anyone would experience plus each person's individual response.

The single most famous example of the technique may be in Russian director Sergei Eisenstein's film *October* in which a shot of an unattended baby carriage rolling down a long outdoor staircase is followed by a close up of an old woman watching, looking alarmed. We cut back to the shot of the unattended carriage gaining speed, bouncing wildly, and when we cut back to the old woman again she is now screaming in horror, her glasses are broken and she has a streak of blood on her face. But we never see what happens to the carriage; the outcome is all the more powerful for enlisting our own imagination to complete the moment of the baby's destruction.

Eisenstein and André Bazin, the French critic, both praised montage for its unprecedented power to stir esthetic responses. It could reach the heart and the mind simultaneously, and because it was in motion it had a force and a poetic effect never known before. They both proclaimed it as a key element in the uniqueness of the new art form. Reading Eisenstein's *Film Form, Film Sense,* and Bazin's *What is Cinema,* I was seduced by their enthusiasm and dissection of cinematic techniques. Yet strangely--and this is how arbitrary memory can be--what I remember most from reading those two books, is not the technique of montage but André Bazin saying that although he loved movies more than any other art form because it was the one which gave him the most esthetic pleasure, nevertheless, the feeling he had coming out of a movie house after seeing a superior film was less than what he felt watching a mediocre production of a play. Why? Because the film was on celluloid and the play was alive on stage. There was nothing between the audience and the actors. It was real people performing in real time in front of your eyes, so despite an inept performance a theatrical production could never lack a human quality that film by its nature didn't possess. It almost disappointed me to read that at the time. I was under the spell of movies and wanted Bazin to exalt *le cinema*, not praise the human power of live theater.

Over the years, I notice that what is human interests and impresses me more than what is technically innovative. The latter flashes its effect and is gone. The human moment, even if I've felt it many times before, can make me stop, feel, and respond again. Which is very different from the often facile effect so easily created by a sudden or unexpected juxtaposition of carefully filmed and edited shots to create a montage, sublime as it may be.

13. Eureka

Writers dabbling with concepts of science, using them as metaphors as I've done here and there, can irritate scientists, but the Triestine novelist Italo Svevo said it was good that writers borrowed ideas from science because even if they got them wrong their creative use of those ideas had the effect of expanding their meaning. For someone like me, who admires Svevo's writing, this defense of misunderstanding strikes the note of unforced charm that I always hear in Svevo's voice. It also reminds me of another author, one known for his innovations as a mystery and horror story writer, who was adept enough at math and science to have made a career from those abilities had his life turned in a different direction. Edgar A. Poe (as he named himself on his books) considered his own best work the "prose poem" portrait of the universe he titled *Eureka*. His intuitions about the universe are impressive, and that's what it is, a vast and inspired intuition--with mistakes, one can see now--

but one that described what would later be called the big bang, among other prescient and accurate insights. (Or at least we think they're accurate today; we might be wrong about the big bang and much else we conventionally think is true about the cosmos.) Poe guessed the universe began from what he called "a unity." He said that light was both a wave and a particle. He explained the reason why, if there are so many stars filling the sky, the night sky wasn't completely white with starlight (because the stars aren't all an equal distance away from us). He said the laws of physics all arise from initial conditions, which he described as an "irradiation" of the unity particle, and that none exists until the unity particle expands. He said the formation of stars and planets was a result of "the condensation of matter." He posited the existence of "dark stars" and that space and duration were one (space/time). Reading these inspired intuitions today in an essay written in the 1840s can make you think there might be some substantial connection between cosmic reality and the human imagination. Poe seemed able to glimpse a universe uncannily close to the one portrayed all these years later by cosmologists using telescopes with high tech image resolution and physicists using particle accelerators and computer simulations. Some of what he described in *Eureka* was off the mark if by that you mean what's been confirmed since by scientific observation. But it was all woven coherently into his projected vision of the origin of the universe. *Eureka* should be better known, not as a curiosity but, as I said, as an example of human intuition providing a glimpse of the cosmos long before we had any tools to look into space, and raising a provocative question about how closely our intuition can divine, or intuit, or mimic, reality. How Poe "knew" these things should give us a kind of sacred respect for the human mind, Poe's in particular.

14. Divine volition

It's all true. But only if qualified by the word almost, spelled all-most. This was my recurrent thought reading Poe's *Eureka*. He wants to refer it all back to spirit, to what he calls Divine Volition, the starting point of the universe in his vision. After proposing this, he goes on to imagine how this cycle repeats itself endlessly. Universes come and go, they are emanations of the Divine Volition. Then Poe goes another step and locates this metaphysical force in us. That's where he loses me. Until then I think his vision is uncannily prescient in the ways it anticipates current cosmology. He seems to have intuited a vision of cosmic existence from only a few facts of Newtonian physics, and if you carve the theology away from the cosmology you are left with his astonishingly accurate vision.

By the time I got near the end of Poe's *Eureka* where he says, "Space and Duration are one," (Einstein said space/time was one), I was thinking about how his assumption that there is a God over all, a divine force that precedes all existence, and pervades it, ultimately and infinitely, makes some of his descriptions of the cosmos more comprehensible than they might otherwise be. This is not a luxury the physicists of today have. Poe has got a conceptual container, a final guarantor of is-ness, that our brains have lost. The God concept Poe refers to has no reality for us. We have lost the ability to believe there must be such a thing. We would be embarrassed to be caught believing it. Poe is not unsophisticated, not at all. He was gifted at mathematics, a subtle as well as a creative thinker, and his ability to grasp physical concepts is far beyond the merely normally intelligent. I think it compares to professional cosmologists, many also gifted with mathematical fluency, because the number of concepts he intuitively understands anticipates so much of contemporary physics. The space-time continuum, for instance, and the expanding universe originating in an initial event blooming with incomparable speed and force from next to nothing, a big bang, is just one of the nearly numberless universes produced by the divinely mysterious whatever-it-is. When you have this final force *God* available to your brain, it puts a frame around the universe you are describing. The question What would there be if there were nothing?, is answered, allowing your mind to separate the universe from the greater is-ness, the ineffable we can't imagine because our brains lack the means. Mine does, anyway. I hit a limit and whatever it is I'm trying to imagine won't take form or shape in my mind. I compare it to infinity. I challenge anyone--any human being, that is--to imagine infinity, to wrap their mind around it. A finite mind can't conceive it except in toy versions like a perpetual trip around a Mobius strip.

What if, however, as a thought experiment, we copy Poe's example and assume there is some ineffable essence that includes our universe--all universes if more exist--and even if we can't say what this indescribable essential quiddity is, can't define or describe it, it's not so farfetched to assume it's existence since something makes our lives and the little we can perceive around us perceptible. The minimum of phenomena we can vouch for is unlikely to be all there is. Not when we keep discovering more all the time. So where does it end? Perhaps it never does but then the unendingness is also beyond our ability to conceive or understand and we can say the one thing we know about it despite its elusiveness is that it includes or in some way frames our own wee universe. Invisibly.

The problem with this is the same problem we've had defining the nature of things since the nineteenth century when Poe's God began to falter--existence has no special role for us. There aren't very many humans who can stay interested in a story which begins by excluding them as significant right from the start. This is the origin of the strong emotional need to imagine and "believe in" a supreme being, an ultimate something.

Stephen Hawking uses the idea of God in his *Brief History of Time*. He isn't literal minded like Richard Dawkins who needs to assert his atheism and then club any believers with it. Instead, Hawking uses God as an idea. When he or it is invoked there's no assumption that a supreme being intervenes in the behavior of the cosmos, and certainly not in the lives of humans, God is just there as the name of a final and ultimate reality, and the curious physicist looks elsewhere for explanations of phenomena. In some cases however Hawking uses God as a concept with a limited, if vast, function. For instance, God could have initiated the big bang that started the existence of the universe. We have no evidence for what happened then, before space and time began, but we can use the name God for a creator, acknowledging the lack of evidence either way, and just borrow the name as a convenience. Other times Hawking finds it useful to cite God as an actor in the cosmos addressing the question about inevitability in the universe. Einstein captured the idea when he asked, "Did God have any choice when he created the universe?" This provides the reader with another reference point, or concept, beyond physics and it helps Hawking help the reader conceive of invisible phenomena which are not easy to understand. This is similar to how Poe uses God to frame his speculations. In Poe's case he simply assumes God exists as a creator and spiritual lord of humankind while Hawking keeps faith out of it. God is the name of an idea. He is using "God" to mean a force acting in hidden, mysterious ways--an explanation for things we can't explain--but he doesn't invest this force with any traits. No intentions, nothing as trivial or absurd as an interest in human beings. It's just a way of naming what we don't know. This is a very skillful technique because it is the same thing humans have done since we began imagining supernatural beings. The difference is Hawking isn't describing God, he is describing the cosmos instead, and God is what he sometimes calls a force we can't observe. It is too elusive, or too something for us to grasp with our limited understanding. This puts his entire cosmology within the greater tradition of theology--in his own way--although the theological portion is narrowly limited and has no relationship with us except an historical one. He shows respect for our urge to believe without letting it distort his physics, which rigorously excludes any myth, anything supernatural.

This idea of using the name of God to represent something cosmological beyond what we know is surprising in Hawking since he was a thorough-going, hard-nosed, evidence-based, twentieth-century physicist. Poe's popular identity is based on his horror stories, not his talent for mathematics and physics, and *Eureka*, his portrait of how the universe began, is only known to literary specialists, and to his fans, so Poe's use of the word God sounds at first like conventional nineteenth century usage. It's not as simple as that.

Poe uses the name "Divine Volition" to identify the force initiating the universe--what we began calling the "big bang" in the twentieth century--and, as he conceives it, it is a poetical evocation of Something created from Nothing. It isn't exactly a reference to the supernatural nor is it the Christian God whom the people in Poe's society believed was the father of Jesus. It's an ineffable Being or Mind and its nature is a mystery which Poe doesn't try to explain. In Hawking's case the name God can mean what we don't know without assuming there is any supernatural being who, if we knew it (which we can't), could explain all we don't know about the cosmos.

15. Dippy

John Brockman's book, *This Idea Must Die*, collects suggestions for ideas that are blocking progress and should be killed, expunged from use. I have my contribution: the idea that there is no God. This assertion, assumption really, by the thinking class is not worthy of science. Where is the evidence that there is no God? It is most obviously the negative which cannot be proved. For thinking people to trumpet their skepticism by saying there is no such thing as a Supreme Being contradicts the principle of skepticism. What it displays most of all is the fallacy they accuse the believers of being guilty of--smugness. Self-assurance, certainty.

Having said this, I still think it's even more silly to believe in a Supreme Being than insisting there is none, especially a supernatural something or other with an interest in us. Even saying it sounds silly. It's the original human vanity, that we should be a deity's chief delight. Nevertheless, when you are reading some work like *Eureka* which uses the concept of God, it feels right and appropriate, as if there's a natural place for it as part of human imagination. In *Eureka,* it is part of Poe's vocabulary, a character in his cosmos.

Richard Feynman used the word "dippy" to describe renormalization, the practice of scrubbing infinity out of mathematical formulations to make them useful and manageable (a process he contributed to himself). Something like that might be appropriate here. We could say using the word God in the way Poe uses it might be called dippy. Yes, it is dippy. It is also useful.

Poe's "Divine Volition" stretches the mental field when thinking about the cosmos. At the same time, the phrase puts a net around the infinite concept and makes it mentally manageable so that even if it's a little dippy the realm one is trying to describe becomes more defined. It's just the way introducing a fudge factor can "renormalize" a calculation that has run off into the incalculable nowhere of infinity, taking all the physicist's useful work with it. Well, nothing is perfect (a phrase that can be read both ways). If our brains were formed in response to life in the savannah so we struggle at times to grasp abstractions, then some kind of aid to thinking may be necessary. Poe's "Divine Volition" may be a fiction and it may leave less trace on "reality" than a neutrino, but if we can use it to represent a force which lies behind the cosmos, beyond the reach of our tools to detect, perhaps we can give it a name and adapt it to our purpose while accepting its dippy, undefinable nature. This arrangement may be only contingent and temporary too, acknowledging that some day we may know more. We will probably never know it all because of our biological limits. That may be why we get these teasing hints about what might be beyond the last mental horizon.

Poe said after he wrote *Eureka* he was ready to die. It meant that much to him to have accomplished it. He does seem to have said many self-dramatizing, self-pitying things during his troubled life, so this may have been no more than an impulse he felt after he put down the pen and not a calmly considered statement. Composing *Eureka* was certainly an intense mental exercise, and he was in very bad shape at the time. In one of those last letters, he wrote to his dead wife Virginia's mother (to whom he was very close) to assure her that he has not been drinking. Yet a few days later he was found drunk in the street and died of what a doctor called "alcoholic poisoning." The detail is sometimes added that when he was found drunk and unconscious he was wearing someone else's clothes. This is a very suggestive note, as well as a very sad one. If you wake up in the street wearing someone else's clothes it's a near certainty that you have not behaved well, and all the more so if you're drunk. That's the disreputable, tawdry, even farcical aspect of Poe's end. (My friend Gordon woke up on a street in Yonkers once wearing no clothes at all. But he was young, a student, a little crazy with youth, and also drunk. The incident got his attention: he learned he had some demons inside him and he had to respect their power to lead him to damage himself through excess.) The serious part about Poe waking up in the wrong clothes was the drinking. It seems to have killed him, perhaps with the help of some other illness. Poe's life and art contain so much lurid material that the way he ended is not even part of his commonly known story. He left us an eviscerated heart pounding under the floor and a live man bricked up in a cellar and several heart-wrenchingly beautiful, doomed heroines--all capped by an omniscient, diabolical raven croaking "Nevermore!" as if a door is slamming forever on hope. All this is more vivid and horrific and original than a dead drunk in the street. (At least he was dressed, whomever's clothes he was wearing.) Poe's imagination outdid his own reality, wretched as it so often was. If he had been able to sit calmly reviewing his own life, the fights he needn't have picked, the opportunities flung away, the way he could offend other people to uphold his own high standards, he'd have seen with another shiver of horror that it all added up to a needless churn of torment and turmoil. Yet what if this was also his own twisted path to *Eureka*, his private hejira into a metaphysical realm, an escape he couldn't have taken without the push from the mad hand at his back. In his darkness he found a light illuminating a vision, and having survived all his self-inflicted wounds for at least that long, to write this in his richly layered and imbricated Victorian prose might have felt as if all the pain had been worth it. Perhaps this vision of the initial singular unity, of dark stars, of the inseparability of space and duration, and the glimpse of what we call quantum mechanics had been revealed to him as a reward for bending his soul toward such a vision so steadfastly. He could have quit a thousand times but never did. And here it was. A man aglow with the knowledge of being. To achieve this was worth anything.

Now he could go get drunk.

I don't say that flippantly. He was alone with these soaring thoughts, he had no immediate means to share them with the world, he lacked even a single wise friend who could appreciate what he had done. He had no one to love him, no woman to put her arms around him, not even someone to shake his hand. He had come to the end of something. Or the peak. The animal mind could make the next choice for him.

16. Depressed

"Are you depressed?"

Gloria, the medical tech, is asking Dad questions.

"Are you depressed?" she has to repeat.

He's 95, he can't hear in one ear, he can't see, his speech is sometimes a garble, and delusions confuse his perception of reality. She knows all this, some of it anyway. It's written in his file. So, how does she think he feels about life? Why is she asking this boneheaded question? She is filling out a form which substitutes for an actual physical exam by a doctor. His face still shows no sign of comprehension. Gloria's own face is full of sweet concern but her musical Colombian accent and very shrill voice--like an oboe with a shattered reed--make a sound Dad has never heard before, to the extent he can hear. She has to repeat everything twice, at least.

"Do you ever feel down?" she is trying again. "Do you have moods?"

Moods. Now the message came through.

"No," Dad said.

She asked again to be sure but Dad was way ahead of her. One trait hasn't changed with his deficits. He knows his own mind and if you ask him the same question twice you'll get the same answer twice. Always.

The whole visit to the doctor was a farce enacted so the clinic could bill Medicare for quarterly visits, several hundred dollars each with no doctor in sight. Perhaps not even in the building. The question was included on one of the standard quizzes they put patients through, mandated by some bureaucrat to give the patient a chance to tell someone if they are being abused at home or by caregivers, or if they are suicidal or have some other emotional malaise. Dad's vascular dementia affects his memory and speech and mixes hallucinations up with the real world, he's legally blind from macular degeneration, he has trouble hearing--didn't I just say all this?--would you be depressed if you had spent your career working as an organic chemist with a highly functioning brain and habitually healthy body and now you couldn't do anything but eat, sleep, and sweep out the carport of your mobile home and bag up the trash? The bureaucracy mocks us with these dumb questions. Yet I realized there was something serious in Dad's answer and it highlighted a significant change in our culture.

For Dad, the answer to the question *Do you have moods?* isn't simply *No, I'm not down or depressed or moody today*. Instead, his answer is global, it encompasses his whole life. "I do not have moods" could be screen-printed on his personal T-shirt. Why has it taken me so long to see this clearly? It hasn't really, I've always known it, I just haven't perceived it with a stoned clarity. This way of seeing him is joined to seeing his will power in action when he's been in a hospital or rehab facility. His steady emotional state, his decisiveness, and his will power combine to impress me with his mental strength. Because he's become physically feeble and dementia has shriveled his short term memory, as well as his ability to express himself in speech, he is now often dependent, like a young child. In this enforced infantile condition, the few abilities left to him stand out in relief. He can no longer express complex thoughts and knows if he tries he'll falter after a few words, unable to find the next word, repeating the last one again, again, then uttering fragments of words, orphaned vocalisms, before all the unspoken words seem to jam up and stop speech entirely. The look of frustration in his glaucous eyes is painful to see and this defeat seems to make him mute for several hours. He sits in a limbo of speechlessness waiting for the words to return. They do not come back, not in anything like the normal pace of conversation. His frustration with this disability is acute, even agonizing, but he is not depressed. Depression is an attitude, a criticism of life, and Dad does not criticize life. He does not complain about anything. There are plenty of things he doesn't like but he doesn't mention them. He complains to his wife, to Mom, when I'm not here, but not to me. With her at times he's a boy, sulky and irritated. But not depressed. He is never depressed.

If he hears a question directed at him in this silence his answer is quick and short, a brevity that is also a kind of revenge on his inability to say more or express all he might be thinking. He knows something is wrong, although he may not be able to remember enough to compare his situation to his own father's, a man who had his first stroke at fifty and lived twenty-three more years with the disabling effects. No one has had the coolness or cruelty to ask Dad if he remembers his father's incapacity and can relate it to his own. What would be the point?

He did say to me once when I was backing the car out of the driveway that it was very hard not to be able to drive. He had a car as a teenager, an old Ford he bought with money from his paper route. Some of the route ran along the shore of Keuka Lake and he delivered papers to those customers using his car. He had spent nearly his whole life with a car, the first generation in history with such mobility. Now he has to wait for Mom to take him where he needs to go. When his frustration comes out in anger, like when he is riding in the car and doesn't know where the car is going, at that moment a neutral observer, like Gloria, might say he has moods. "You're going the wrong way!" he shouted at me from the back seat. He never shouted in his life. His own father shouted and it embarrassed him. A self-possessed man didn't do that.

Dad's loss of independence came in installments. First he couldn't drive. Then he couldn't take a walk by himself. When I called, Mom said, "Tell your father why he can't walk to Walmart's." The big store backed up to the mobile home park and was only a half mile away. But Florida is not designed for pedestrians and I told him he might get hurt or run over because he couldn't see, or he might get lost. He wasn't worried, he said, arguing as if he was still whole. He wouldn't listen. Finally, frustrated myself, I said, "You've had a stroke, Dad." "I *hate* that word!" he shouted, more vehement than he'd ever sounded. To get his attention I'd gone too far; I'd said the cruelest thing I could say.

He did complain in the rehab center, eventually, probably because of the drugs he was taking, which altered his personality (he had moods!). They gave him quetiapine, an anti-psychotic. He wasn't psychotic, he'd had a stroke! The drug was to calm him, control him. It made him hallucinate. We only found out what they were doing later. It made me angry enough to do something violent. This is my *father* you're drugging! (I have moods.)

Dad wasn't the kind of man who had heroes--there's that self-possessed quality again--but consciously or not he modeled himself on real or fictional men who were steady, stoic, and suffered in silence. That suited his nature and was also the masculine ideal of the culture he was raised in. It was different from his father who, especially after his own stroke, was volatile. He cried easily, he got into arguments with his retarded daughter, rows noisy enough to ruin dinner, and in one of the very few incidents from his youth Dad described to me, his ex-Baptist preacher father delivered an angry, impromptu sermon denouncing the alcohol being served in the fraternity house at the college Dad was attending. Not only was the content of his lecture humiliating for Dad to endure, but his father delivered his tirade in his incomprehensible speech. So besides losing his temper, he made himself ridiculous, an object of laughter and mockery. I doubt any of his frat brothers blamed Dad for his father's outburst. Instead they probably sympathized with him, all the more so because Dad had no taste for alcohol himself, but would never have presumed to tell anyone else what to do.

17. Charmless and naked

My own bout with depression, the kind akin to what Gloria had asked Dad if he ever experienced, was so wrenching and debilitating I really don't want to exhume it (anymore than Dad wanted to be reminded he'd had a stroke). I'm not sure I trust myself to portray that lost soul either, although one theory says whatever I say must be a version of *some* truth authentic to the experience, no matter how slender or mistaken, so just say it and see what comes out. And to digress for a moment, in the hope that the subject will melt or vaporize in the meantime, this technique is the one Montaigne used--just say it and don't edit or correct yourself. The writing may suffer for lack of second thoughts and revision, but the portrait will be more authentic (there's that word again).

To put it melodramatically, depression is a descent into hell. The hell is not other people, as in the oft-quoted quip, the hell is yourself. It's the self you despise, the one holding you prisoner in its contempt. How you find yourself there is always a mystery because it's a place you can locate only once you have fallen into it. You don't do anything to get there and once there you can't do anything to get out. You often deny the hell you are in. It's embarrassing to abdicate your own life. It's damning yourself, treating the gift with contempt. This is because the world we all inhabit, the world around us, is, objectively, not a gift but a version of hell. Life sucks, that's a non-trivial fact. It's a meaningless, pointless, futile treadmill where all emotion is some form of ennui. That is the sole truth saturating human existence and anyone saying anything different is under the spell of a sentimental delusion, denying the truth, the obvious truth so they don't have to face it.

On and on this cycle of solipsistic reinforcement goes and the longer it continues the harder it is to break free. That this state of mind is a disease, a mental illness and not simply an attitude, should be apparent. It may be a delusion that life is worth living against the evidence of its absurd futility but whether it's a delusion or not it's one we all need. Delusion may not be the right word. It sounds like the vocabulary of the depressed, not of a normally healthy temperament. Good health is not a delusion. What keeps us going is physical energy. It is what denies, or alters, the meaning of the word delusion by infusing us with an emotion. The emotion is the desire to live, to be alive, to experience our sojourn on earth in the company of others like us, breathing in, breathing out, and feeling all the rest of the physical and mental sensations accessible to a healthy mind and body. For this we're willing to endure drudgery and pain, according to the limits life trains us to accept. In this way, our sojourn is indeed a negotiation. One thing we have to negotiate is the way normal society and particularly governments and their agents do things that are sick. They approve of immoral, demented behavior to keep the rich rich and those in power in power. Those who wield wealth and power are often the healthiest, most vigorous, least depressed people among us. This is a fact we must all live with. The ideal ethical world we may have imagined or been taught to believe in as children doesn't exist. For an ethically aware adult this is the subject of an ongoing, endless negotiation between the ideal and the real.

Our depressed person is not interested in negotiation, they are thinking in absolutes. In absolute terms, the depressed person is right, life is an amoral scam, a trick evolution plays on us to get us to convey our genes into the next generation. That's the pathetic, pitiful deal we make with life. The depressed person, his sense of humor lost, his energy drained, his spirit wan, is left with only a mordant residue, often bearing the burden of a crushed ideal on his back. What can he find to celebrate in the progress of the thinking biped, so proud of his brain, strutting his way through life on the strings of an invisible puppeteer, ignoring whomever he might be trampling underfoot. The healthy person merely laughs and carries on, accepting this condition as if it was his right. It's enough to make you depressed. What should we do? Copy the doctor who ministers to the unwell without getting sick? To one who's depressed it feels like another failure, more evidence that, honestly and ethically understood, life sucks. If you were healthy, you wouldn't care.

I liked the shrink. He didn't look like a shrink. He was a young man, husky and bluff, commonsensible, direct. No jargon. I sat in a chair and talked. Talking was something I was good at. Or thought so in a flagrant and no doubt foolish and transparent show of vanity. After two sessions the doctor suggested I lie down on the couch. I did it reluctantly and was right to be wary. Talking to the ceiling was weird. I felt as if I was floating in a sensory deprivation tank with nothing to hold on to. In this state I might say anything. I was probably afraid most of all of bursting into tears. The emotion underlying depression is self-pity and to prevent the onrush of this humiliating feeling I needed my wits to carry my end of my argument against the world. Once I lost justification for condemning Life As I Knew It, I was naked and exposed. Charmless and unlikeable.

What I remember, or think I remember, about the session floating on the couch (not so different from gliding through space on a pyramid; no, no--totally different), was that my reluctant monologue lit on the subject of Dad. Given how important he was in my life and how much our fractured relationship hurt us both, this was probably inevitable. It must have gratified the doctor to hear this. Instead of green, rambling, quasi-intellectual fatuous opinions, he was hearing at least a sampling of raw psychic material spilling out of his patient. My father, I might have been saying in some form, was an impossible model to live up to. He was the kind of person who never did anything wrong, not through vigilant self-control but naturally, because it suited him. His appetites were modest, his hobbies were his family, taking care of his house, going to church and serving on committees. He had a good job as a research scientist which he mostly enjoyed. He was happily married, although that would change with the turbulent sixties and seventies as his wife's dissatisfaction with a constricted life morphed into her own mental illness. (Was there a link between her illness and mine? Not really, but then, yes, of course. How could there not be?) Dad's flaw, which ran through his life like a vein of quartz in a mine, was a certain rigidity, an emotional narrowness, and provincialism. The latter he could see in himself and compensate for or correct at times. In theory, he could also see and to some degree enjoy the range of human dispositions. But when it came down to the private man, the son of his own father whom he was helpless not to at least partly resemble, he was unable to see why anyone should be different from him. His own limitations and taciturnity were balanced by his modesty, decency and sense of humor. Intelligent and trained as a scientist, he was quick to sense what it was in any situation that he didn't know. The picture of life in his mind made logical sense, even if he knew no one's emotional life was logical. That's not how we're made. But we can mold ourselves, control our actions and do the right thing, and that is part of building character.

From the doctor's point of view sitting by the end of the couch his patient is floating on, the case must have seemed self-evident. The familiar conflict between a traditional father and a rebellious, malcontent son struggling to fit somewhere in the adult world. But I have to interrupt myself here. I don't like where this is going, just as I was uncomfortable lying on the couch telling the shrink whatever I was telling him. This was more than fifty years ago. My father is dead and long before he died we resolved our conflicts and he showed great generosity and an effort to understand not just me but all his children, and his wife too, a more problematic situation. It's true that people who knew me sometimes commented on a steely quality in the man. "I'd hate to go up against your father," said my friend Carl, describing the stern glint that could appear in his eyes. That is just what I was doing, going up against him. But I don't think Dad much liked that trait in himself, either. He inherited his version of it from his own father who grew up in a time when the role of paterfamilias was supported unquestioningly and almost universally in our culture. Embodying and enforcing a code of behavior was a standard a father had to come up to. Dad used to joke about it, but teasing himself was also a way of paying tribute to the recognized role. Father knows best and he decides what's what. It was a role that could imprison both fathers and sons and their wives and mothers too. It led to willfulness, to fighting and grief. These men who had to uphold a standard they often felt they were failing themselves had their own struggles with failure and humiliation. Dad had few disappointments in his life because it was, overall, successful. Except when he was pushed hard, he was not arbitrary, and didn't take his own frustrations out on anyone else. He suffered them silently, arguing with himself, I assume, that he was mostly a lucky man and it was unseemly to complain. In most roles in his life he wasn't a tough guy at all but decent, genteel and forgiving, the kind who never hurt anyone.

My conflict with my father did not make either of us a better man or make us closer when we got through it and could appreciate each other again. We were so close when I was a child that simply to return to an adult version of that relationship was more than enough. Both of us had been helpless in our roles. We got through it and not so long later when my brief marriage went out like a candle in a puff of wind, he didn't say, "See, you're still screwing up," but offered sympathy and support.

When the fifty minute hour was up, the doctor promptly but politely stopped the flow of words (I hadn't burst into tears, but came close), and said, "I think we made some progress today," something he hadn't said after our more social sessions when I faced him in a chair. The free-floating sensory deprivation mode had worked. It may have worked too well. I was already uncomfortable with these weekly visits to the psychiatrist, and not the least reason was that Dad was paying for them. They weren't cheap. He had even been driving me to the office and picking me up again, and although he said nothing critical I knew he only went along with the contemporary taste for psychological self-investigation because the real doctor, the internist, had recommended it and was a friend of the family. So the sessions added another layer of something insoluble to our relationship.

A few days after the third meeting with the shrink, the one on the couch, I threw my small bag of personal stuff in the trunk of Fox's old powder blue Caddy and we took off west. It rang in my ears only briefly, the doctor saying we had made progress in our last session. No, I saw no way forward with him. My role was too passive. If I was going to plumb myself and the particular slice of reality I found myself trapped in, I was going to do it on my own.

How much yet how little there is to know. The chain of connection from grandfather to father to son is only three people long. Behind the first is a blank, a farmer from Pennsylvania. It's curious that in the long string of humanity running all the way back to Adam (as Twain would say), the only ones with any identity are the last three. It's nearly the same thing for everyone. The facts and details that fix who we are, without catching anything of the essence, all vanish as swiftly and completely as last year's compost, and in what seems an instant we are all no more than the children of one or another Farmer Blank from Pennsylvania.

18. Pathology

This collage shouldn't become an account of mental derangement or emotional sickness. It's true Eli went through such a period and it's true that we live in an era so deranged at times by changes in how we live, how we behave, what we believe, what we know, and what we don't know we don't know, that a pathological portrait is the most apt one. There is a madness in us, an inability to handle, let alone control, all the powers and baffling ironies we've unleashed starting with the Enlightenment, then the Industrial Revolution, whenever we agree to say those changemaking epochs began. Henry Adams used the dynamo as the symbol of this transformation. Its power was blunt and mindless yet superior to art and religion. What could you say to that? Nothing, and since then it's only gotten worse. We've got to deal with it, there's no choice, and now we're threatening our own civilization and even in some ways nature itself, fiddling while the polar ice caps melt, wildfires rage, coral reefs die, the oceans warm and acidify blah blah blah which makes any individual's personal anguish...it puts it into perspective. On top of that, and more fundamentally, I'm not a whiner or complainer but a positive, upbeat, humorous, ironic, faultfinding, pessimistic, doomsayer who does not believe any god will save us and The End is just up ahead.

On the other hand, look at all the truly fine science books being written and published just in the past fifty years. This outpouring of smart, curious minds studying nature is unprecedented in human history. There are far more than a single person could remember. Has anyone stopped to take note of that fact? The voice of sanity, reason and intelligence is available almost everywhere and can be posed against the madness that seems at the same time to be bleeding out of the body politick, staining everything a furious red. How can all these sane voices of science be reciting their discoveries and observations without having some saving effect on human fate? Do you think it's possible they could be ignored, even silenced by some destructive global catastrophe? That all these thoughtful works of knowledge could come to nothing? Could a pyramid crumble to dust and blow away?

19. Cenotaph

Here lies the human race which lacked a key survival trait.

20. The Abingdon Bible Concordance

It was a big deal when this hefty book was brought into the house. Maybe I only got that impression because it was such a large book with a name I didn't understand and when Mom told me what "concordance" meant she might have sounded a little awed herself. What were they doing with a scholarly index of names and words in the Bible? Were they going to use it preparing Sunday school lessons, which each of them taught at some point? The book wasn't opened very often. It may have had a more symbolic than a practical purpose. I can imagine Dad wanting to be sure he got a name or reference right and the Concordance could help there, so you knew Abram was the same person as Abraham, to pick an example they both would have known without looking it up. Religion was important to both of them but neither was pious. Church-going was as much a social event as an act of faith. Many of their friends were people they met at church and they were comfortable in the atmosphere of decent folk who shared their values.. So buying the book--and it was expensive--may have been spurred by a specific incident, a need to find something out. This was long before you could look things up on the internet. Stumbling across a name like Melchizedek, for instance, which is obscure but resonates with meaning as an example and prefigurement of immortality, might have been enough motivation to spend the money on the Concordance. Recalling the name Melchizedek reminds me that toward the end of his life Dad told his daughter he no longer felt he could believe that we would enjoy some kind of spiritual afterlife. The winds had shifted; they blew from the regions of science, not religion, and he was a scientist himself. Besides, many things change in old age, often surprising the one who is aging.

Mostly the big book sat on the shelf radiating a kind of stolid, self-confident power which I always felt when I noticed it. Since I don't remember opening it to look something up, that was my relationship with it in the house--a fellow resident, a very quiet one, who contributed its presence and little more. Yet that little was significant and similar to how I felt about many of the books in the house. They were silent companions. I didn't read most of them and when I did they often put me to sleep. There was the big, heavily bound ochre colored *Stories from the Bible* with gilding on the edge of the pages. Its imposing look and feel were appropriate because its stories were only a little more friendly to read than the ones in the Bible itself. Another constant companion was Mom's dictionary from college, the *Winston Collegiate Dictionary*, also a fat book, with a black hard cover and indentations with the letters of the alphabet stamped on them that just fit the end of a finger so you could quickly find the section your mysterious word was hiding in. It wasn't as big as the massive dictionary that came with the *Encyclopedia Britannica, Jr.* set of books, so it was handier, but it was distracting to use because it had been defaced. Not by Mom, but by me. Throughout the book, in the open spaces between the definitions of words, in the margins, at the top and bottom of pages, there were squiggles in blue and green ballpoint ink. Sometimes there were shapes, like clouds, or slinky-like circles, something like the writing exercises we did later in school. These primitive, illiterate scrawls were something between writing and drawing, a crude expression of the raw urge to do exactly that--to express something. But what? The urge had no art, no language, nothing to say. Nothing except that raw urge which can be taken to stand for this force within us to give voice or coherence to our amazement or puzzlement at being here. Or if not amazement then frustration, because when I look at these squiggles now I think I can feel, remember feeling, a thwarted sense that this was the best I could do, or all I could do, and what I felt seeing them was their inadequacy, their failure to convey anything at all. In that sense, someone with a wry, even a bitter, sense of humor--a writer, for instance--might say they were an accurate representation of the experience of writing. A futile activity which leaves one feeling inadequate and frustrated.

Maybe it's also the emotion felt by the Neanderthal man or woman who was recently given credit by archeologists for the red and black lines, marks, and symbols painted in the wall of a cave in La Pasiega, Spain, more than 64,000 years ago. One painting showed the outline of a hand, a signature of the artist, we could say. We could say all kinds of things because the marks are simple enough to allow many interpretations, as many as used to be extracted from Rorschach inkblot tests. For me what's compelling is the blunt evidence of the urge to leave a mark. Because it required an effort to do it, and it also required forethought, even planning. For example, how did the artist see his or her way into the cave? Did they have fire, some kind of torch? Were there crevices where sunlight reached down into the cave? The scientists don't say. All we know of the original artists is their kinship with us through the urge to say something, do something, to indicate a consciousness that they're alive and want someone to know it. Someone back then was trying to say the same thing Faulkner said in 1950 when the Nobel committee honored him with a prize for making his own marks in red and black on the wall of his cave. Those marks, interpreted by other humans, impressed the committee with the beauty and meaning of language and the common humanity of us all.

And what happened to the Abingdon Bible Concordance? It went into the dumpster in the driveway when we prepped Mom's house for sale and had to do something with all the contents. No one in the family wanted it. Not even me. I do want it now, but I was in a ruthless mood then, as Philip said was required of us if we were going to look through a lifetime of acquisitions and sort out the very few things to be salvaged. I didn't take it because I knew that after this moment of decision, take it or toss it, I would never look at it again. For the next several days as we cleared out the house and filled up the dumpster, I kept stepping over the Concordance, as if it was finding ways to stay in sight, to remind me of my perverse streak of contempt for family history, counterbalanced by a paralyzing respect for all histories, perhaps even primarily the history of the Bible.

21. Credo

Martin's question keeps coming back to me. It was when he and Anna were here for dinner and we got into one of those conversations where everyone is talking a little too loudly and eagerly about what they believe. Or don't believe. We all share a similar skeptical humanist attitude but when people are asked whether they believe there is any kind of spiritual dimension to life some unexpected things can come out. Not so likely with this small group because we know each other too well; however, on this night there was a slight variation, probably because Martin's cancer has made him thoughtful in a fresh way. The treatment is going well and no one is warning him he might die soon, but still, one can't avoid mortal thoughts when you have a tumor on your liver. So there were some noisy wisecracks about religion and Martin expressed his usual sardonic view of the whole thing as vain and wishful thinking, fueled by his opinion that the culture of believers was an alien one for someone with his background. To him it had always seemed smug and exclusive, a club of the ruling class with nothing spiritual in it at all. Nevertheless, he wanted to hear my opinion. (Interesting that I use the word opinion since how could a mere opinion affect the spiritual? Does opinion have the power to create or cancel the spiritual?) Before I answered I remembered that night long ago with Martin and Anna when we were all full of bourbon and I went on about the holy spirit of man (or something like that) and afterwards Martin said my preaching hadn't made him believe but if anything could make him believe what I said would have done it. I wondered if he was hoping for another performance like that, for its entertainment value or old time's sake. But I'm too old to play the role now. I can't produce so much enthusiasm or conviction, and as a symptom of that change I prefer silence to a boozy torrent of words. I also knew what could happen when one tossed a serious thought on the dinner table along with the plates and glasses and empty bottles. Yet lately I've been thinking Martin's question is one I should really try to answer. Not with another overwrought monologue celebrating that dubious numinosity, the holy spirit of man, but to see if I could put into words the feeling I have about life on planet earth--speaking as one of our species--and how it relates to the cosmos. (What an invitation to dippiness it is when I write it! Yet I can't squelch a mental phenomenon like the numinous; maybe I should call it contingent instead of dubious, contingent on belief itself). Some of what I think or feel has gone into pieces I've written elsewhere in verse but as obvious as those observations seem to me I notice that people need a pointblank statement in prose. Something almost like a confession of faith, as deadly a form of literary composition as exists in any language. What they don't want is ambiguity, the very quality that emerges to haze any depiction of the spirit. That itself is a problem since what one is trying to describe isn't as literal as what people want to satisfy their yearning for meaning. I know because I want it too.

Once before I sat down to try to express what I believed. That time I failed completely. I was about twenty years old, sitting at a desk in my bedroom, the desk my grandfather had used to write his sermons on. I didn't get much farther than writing

"Credo" across the top of a blank page. All I had was the urge. (Not unrelated to the urge that made me deface Mom's dictionary and led our Neanderthal cousins to mark the walls in a Spanish cave.) I didn't know what to say. The sense that I knew, or believed, something yet could barely come up with a single sentence to express it was very frustrating, and humbling. A Zen devotee might find this quite amusing and if I was offended by his laughter at my young, touchy age, he might have said to reassure me, "A blank page is not a bad start. Why not contemplate that for a while." (Like for fifty years?) Well, I did, but with no encouragement from anyone, including myself, the experience inoculated me against any more rash, head-on attempts for years afterward. The Zen devotee might have added to reassure me now, "A blank page might be an even better place to end up." If I'm honest about it, I have to admit I probably contemplated the blank page for many years without feeling I could improve on it. I still feel it's a challenge to improve on it and the wisest way to begin is to acknowledge that one's attempt will fail and the most one can hope for is not to offend silence too loudly. If I could, if I knew it, I'd inscribe in a pleasing calligraphy the Japanese hieroglyph for silence at the top of a page and contemplate that, hoping the beauty of the brushstrokes was an answer more eloquent than any I could give in plain type. The idea that the brushstrokes representing a word can have their own beauty independent of meaning, is one I envy, but it's not one an individual can create out of nothing. It's an artifact that needs a whole culture to nurture it to life, not just a single person. (I've also been told that the hieroglyph's meaning has nothing to do with the esthetics of calligraphy. For my literal mind that was a little disappointing, but I can see why it's so.)

Going back to what I was trying to say before I interrupted myself, the best thing I can say, so I don't smudge the blank page too indelibly, is that meaning is up to humans to create; it's up to us to put something on the blank page, both for our individual selves and for the species. If there are any sentient beings that exist between us and a Supreme One some people can still imagine exists, their existence is not affected by whether we believe in them or not. As for the holy spirit of man--we know it exists because at times we can feel it, or hear it, since it is probably most manifest in the music we make. What is that emotion we feel listening to a choir of gospel singers? It may be merely the most stirring expression of yearning a human being can make, but if we were scientists measuring the response of *Homo sapiens*' heart and brain to that sound, and we found no emotion could exceed it, we would think we had found *something*. It wouldn't matter what we named it. Except that if we did give it a name, that name would soon be tarnished by too casual or vulgar use, might even become a curse, such is our need to express the frustration we feel when life fails us. Or we fail it. But at its best, at our best, at the other end of the scale, we may hear the choir making a sound we might call an invitation to all the rest of the species, to any living thing, to feel the experience of life and sentience in a cold, dark universe. (If I leave this and come back later to read it again, I'll probably delete it.)

Why not leave this attempt at a *Credo* right there in the heart and voice of the singer, or leave the moment of transcendence in the illuminated mind it is passing through, which may not be ours but the organ of someone more alive, more sentient, with denser and finer neuronal connections, rather than risk diminishing sublimity with words which can only invite mockery and irony, denying the spirit its ambition, leaving it with only a sigh.

22. What could have been

Antarctica is melting, or to be more specific, a Mexico-sized glacier on Antarctica is melting, a possibility the scientists thought wouldn't happen for years; instead, they find evidence it's already happening. That means a much higher rise in sea level than predicted before. The mind reels back, reeling from imagined calamities to the question--how did all this begin? With the industrial age in the nineteenth century, some say, then it sped up in the twentieth. The changes have come so fast, too fast for many to adjust their view of planet earth and our place on it. Was there a time we could have changed our trajectory? My mind floated over the years and stopped at World War 2. That was the event, the worldwide event, which shifted balance toward us, the U.S., and our expanding, culture which soon became profligate (partly made possible in the nineteen-fifties, -sixties and -seventies by a high tax rate on the rich). Once again, look how fast everything happened. In 1943, the middle of the war when I was born, my parents lived in a cold water flat in Columbus, Ohio. No, not even a cold water flat, a *no* water flat. To get water they had to go out in the hall, walk down to the bathroom, and fill a kettle or bucket from the faucet in the sink, then take it back to their apartment and heat it on the stove. There was no running water in their kitchen. Heat was provided by a coal furnace in the cellar which left a residue of soot on all surfaces, in the air, on the floor, in your lungs. Dad was a grad student doing war research at Ohio State University, living on a slender salary in conditions not much different from many other "middle class" young people. This was more than seventy years ago. A single person's lifetime. When the war ended two years later in 1945, with America victorious and so much of the rest of the world--both winners and losers--flat with exhaustion, could we have imposed some kind of new world order in which people produced and consumed wisely, using only what they needed, without fouling the air and water of the home we all shared? Could that have been some kind of new start for mankind with all of us motivated to share resources based on our common humanity? Dream on, bunkie. The suggestion is a joke. It could only be made by someone who has never experienced life as a human being themselves. The members of our species, each seething with their own urgent needs, could never moderate their appetites so wisely or equitably. Whatever one has, one wants more. Even those with enough to hoard or waste--the rich--want more. (Those high tax rates on the rich are already history.) That's the nature of our organism. The great war raging over the planet in 1943 was a product of that rampant nationalism and what might have looked like a chance for a completely new start to life on earth at the end of that universal struggle was nothing of the kind. It's so obvious, it doesn't need to be said, not unless one is musing about something as impractical as the question I was asking myself, could things have really been different? In the late nineteen forties the world was churning with almost as much desperate urgency as when everyone was making war on each other. In the blink of an eye, our ally became our chief enemy, a new "cold" war took the place of the hot one and began to dominate our attention. The only people worried about how we used earth's resources were a small number of rather quaint outdoor enthusiasts called conservationists.

From the peak of a pyramid in space, the temperature of the air or how high the oceans are on earth is inconsequential. If you take a long enough view, most any problem can be diminished to nothing, just as if you get close enough to any particular problem, you can't see anything else. As serious as global warming is and as much as it will affect everyone, and as close as all of us are to its consequences, it hasn't motivated real action because it conflicts with the race to get more for ourselves. Nothing can change that. Nothing but a calamity that exacts a tax on all of us--unequally, as usual--so the ones at the bottom pay dearly, even fatally, and we pay proportionately less moving up the ladder of wealth. Everyone knows this. There is no kind of leadership which can do anything about it. Not yet, not significantly, not in proportion to the approaching calamities. Soon there will be a name for a climate-related calamity the way "9/11" stands for the emergence of terrorists who want to bring down western civilization, or "Chernobyl" represents a runaway nuclear reactor. It will be the name of a place, a city or country or region which experienced flooding or drought or fire or something we can't imagine yet and that name will be a marker, a secular holy day in the chronology of our species, a day of regret and atonement. Too late, of course. Then we will say, "What could we have been thinking, to let this happen?" But those words will arise from a feeling we don't yet have, from an experience, a new fact, which we can't name yet. As inevitable as this is, as visible as it will soon be to all, it is also something we can't quite imagine. Not yet.

Or maybe Act 1 is playing itself out now before our eyes and we can't see what we're looking at.

23. A sane person

When people are behaving insanely--and in our case it's an entire civilization which lacks an ant-like common mind so is unable to save itself from its own excesses--it makes you value the trait of sanity all the more, along with its cousin moderation. In his *Essays*, Montaigne's self-portrait is telling us the way the world looks to him and what he is like as one of its residents. In many instances he carries this out to it's logical conclusion, asking what should be our practice in whatever situation he's describing. This leads to my subject of sanity because his own opinions and behavior seem quite sane and when they differ from the norm he's aware of it, describing how and why he deviates from what's typical for one of his species and culture. He often doesn't defend his own quirks, but only describes them. For instance when he says he would rather not know if someone is overcharging him for work done on his estate because he doesn't want the hassle of an argument.

He knows he is occasionally cheated but the amount is not great enough to compensate for how upset it would make him to insist on squaring accounts more strictly. Besides the money, this policy also loses him some respect among the workers, he assumes, and is already resigned to being regarded by them as what we might call a klutz. Montaigne admits he doesn't know how all kinds of ordinary tools, devices, and appliances or common daily procedures work, despite the many times as he's seem them in use. This, a reader thinks, must be either indifference or a lack of skill with his hands. (Perhaps the indifference was a result of his ineptness; he doesn't try to explain his deficiency.) In describing these traits he left me with the impression he floats over daily life in a kind of preoccupied state of mind, and elsewhere he admits he is susceptible to daydreaming, to becoming bored in conversation except with a lively companion, to dozing off in company if the talk is merely polite. You can imagine that the workers and servants at the Montaigne estate are amused by their ineffectual landlord and employer, and as proof Montaigne says there are a number of stories told about him, anecdotes cherished among his hired hands as hilarious examples of his fecklessness. Now the picture really takes shape: Do you mean, Michel, that people tell stories about you? And laugh out loud? So if I were to suggest to your contemporaries that Monsieur de Montaigne is a good example of sanity, they might think that was hilarious too. Instead of wise, they might have thought of him as an airhead. But you also notice that the worst evidence of his incompetence comes from Montaigne himself. He is a credible witness, either for or against something, including himself, and that is also a mark of sanity. He seems to see the world as it is and is comfortable in his own skin, even when it provokes laughter. Instead of a mystery, he's an open book and that book is his own, freely composed. He tells the reader he is writing his self-portrait so his relatives and friends may know who he was when he's gone. He has no interest in a wider audience. If this sounds disingenuous--and I think it is--it may also be an author's modesty. Whatever the reason, his descriptions and opinions never attempt to seduce or ingratiate. He just serves them up as they occur to him.

None of this is getting me closer to what brought a sixteenth century writer into view from the pyramid. What secures Montaigne a place here was his saying of his father that he was "the best father ever." The theme of fathers and sons is one I can't get away from. When I hear Montaigne's quotation repeating itself in my mind, it occurs to me that the best father ever requires a very good son. Because of some tricks of nature, I wasn't as good a son as I should have been. My father could have been even better than he was, which was often exemplary, but some other person than the son he would have chosen for himself was struggling to emerge from my make-up, and to acknowledge or to fulfill him--to try to set him free--I had to push my father away. Finally he rejected me; then, in a silent reconciliation carried on by signs and gestures, he seemed to forgive both me and himself.

Montaigne was most fortunate in never having to go through any wrenching growing pains with his "best father ever."

24. Noah's nakedness

I'm sitting in a chair in a patient's room at Morton Plant Hospital in Clearwater, Florida. There isn't much extra space in the room, especially with a nurse and technician standing over the bed, which is why my view of my father at the moment includes his genitals. I'm trying to remember the name of the character in the Old Testament who was punished for looking at his father's nakedness. A glance into Genesis recalls the story. It was Noah's son Ham who saw his father naked. Noah was drunk and Ham went into his tent and saw his father "uncovered." When Noah learned this, he cursed Ham's son Canaan. As often happens with Biblical stories this one is laden with meaning, but it is also somewhat incoherent and has to be explained to make sense. Then it has to be further interpreted to do more than merely make sense, which is never a priority in Biblical tales anyway. The priority is serving its purpose in the history and mythology the ancient Hebrews are creating about themselves. In this case the purpose is putting a curse on the Canaanites, the long time rivals of the Israelites--and here we are today with their descendants, the Palestinians, people at least partially sharing the Canaanite lineage, still locked in conflict with the children of Israel. None of this politics is relevant here except that a reason for Ham's heirs to be cursed was needed and the oral tradition which conceived and passed down this story used breaking a taboo as the crux so the message would be forceful, and adhere permanently to the Israelites' shared memory. Everyone in these tribal cultures knew that looking at your father's nakedness was forbidden, and sitting at the foot of the bed, I can feel that. I shouldn't be seeing my father's male potency so exposed and helpless. I may only feel a passing shadow of the taboo, but it's noticeable. It makes me uncomfortable in some primal way. I am suddenly not a child of the twentieth century but a common human protoplasm trapped in the peristaltic passage of my species through time. A taboo can blight me as certainly as it did Ham, even if all this happened by accident, and fleetingly. Only a glimpse of Dad's nakedness was exposed to me before Luis, the affable, oversized nurse, moved the plunger he was using to irrigate Dad's bladder, and obscured his organs again.

Like everything else, taboos are put in perspective by Dad's condition. He is ninety-six, he has an infection making him sick enough for his condition to have been described as "dire" by the doctor, and now he has blood clots obstructing his bladder. I'm watching Luis because I want to see if he is going to be able to clear the clots. He does it by injecting water into the catheter inserted into Dad's urethra, then uses the plunger to pull the fluid back out with enough force to break up the clots and empty the bladder. Luis has been explaining to me that the catheter has two small holes in it so he can inject on one side and drain on the other. The procedure takes a while and he's not at all bothered by my watching him and asking questions. His size and matter-of-fact confidence make me feel that if this can be done, he can do it. Yet simple as the procedure is, it's compromised, or limited, by trying to fix something inside a body from the outside. Luis knows his patient is ninety-six and in "dire" condition, that he has dementia and therefore probably has only a vague idea of what is being done to him. Whatever he feels, Dad hasn't made any protest, and Luis works with a professional steadiness.

This morning we had the DNR conference with a doctor, a nurse, Mom and my sister Jean. The nurse had already told us that if Dad's heart stopped and they performed CPR on him, the chest compressions might break his breastbone or ribs. She made it sound as if this was more likely than not, and not a good idea, even though it was the patient's decision. Or rather, because Dad is too ill to understand the choice, his proxy, Mom, would make it for him. I'm the back up proxy and way ahead of the nurse's recommendation. I don't want him to suffer and since Dad has already signed a health proxy back when he was "of sound mind," there should be no problem. My sister agrees but Mom is the one who has to sign the DNR and she is worried in her wary, anxious, almost childlike way that someone is trying, subtly, without saying so, to kill her husband. Just as they might do to her if she were sick and they wanted permission not to revive her. Her underlying fear is that she, or one of her family, will be buried alive, so whenever she hears the word "revive" or "resuscitate" a gong rings in her mind and a wave of fear comes over her. We finally persuade her to sign the order with the argument that it's Dad's own preference, clearly expressed more than once, but it's like trying to coax someone who can't swim to jump off a diving board. She only agrees because Jean tells her she can change her mind and rescind the order any time.

The way the issue plays out in the family mimics the larger debate. Dad spent his working life as a scientist and believes doctors tell the truth as much as they know it. He trusts them, and the body of knowledge they've been trained in is based on the same assumptions as the one he learned. Mom spent her life as a housewife and school teacher, and although she is educated she doesn't understand the scientific method or give it any special respect. She trusts her intuition, if she trusts anything, and since her bout of mental illness in middle age a number of shifting delusions have colonized her mind, so her opinions are a mix of clear and often perceptive insights (especially about human nature) confounded with notions and fantasies that might come from anywhere including television, a newspaper or magazine article, or rumors passed around casually among people she knows. Or doesn't know. Besides these ingredients there's the most potent factor, her own imagination. She is a kind and loving person with no malice toward anyone, but she seems aware that these positive traits can make her vulnerable at times, so for this and perhaps for other complex reasons she can also be suspicious and fearful, often of threats no one else senses.

As things turned out, we didn't have to invoke the DNR directive. Dad survived bladder irrigation and the dire infection, but only for another several weeks. He was sent home from the hospital, then almost immediately moved to a care facility where he died peacefully without any lapse into medical "unresponsiveness." Mom was spared feeling that her decision had "killed" him, or that we had done the wrong thing. When she said goodbye to him as he was being wheeled out of the room to go to the funeral home at the ghostly hour of three in the morning, she said, "Bless your heart, dear. Be happy, wherever you are. We love you." Her sentiments, and the hint of the supernatural, were entirely in character. She wasn't so much upset by the death of the husband she'd lived with for seventy-four years, as she was bewildered. What was going to happen now?

25. Lowry's conundrum

In trying to express the meaning of some particular thing, people sometimes make analogies with scientific concepts. As I have mentioned before, this is what Italo Svevo was referring to when he said it was all right if writers did that even when they misunderstood the science because it expanded the meaning of the concept. I think it would be better if people got the idea right but that is not always going to happen, especially as science learns more and the ideas that get into public circulation become more abstruse and harder for laymen to understand. Yet sometimes just the name of the concept is enough for some notion to be used to mean something in the non-scientific world. It's like the way the Higgs boson picks up mass as it flies through the--ah, see how I'm doing it again right now, and so glibly, as if I really understood anything about the behavior of subatomic particles. I can't quite stop myself from doing it, wrong as it may be. The best concepts to cite may be the ones incompletely understood by the scientists themselves because they have been so newly discovered. There is also an appeal in sidling up to the unknown because of the power mystery has over all of us. In the end, however, this is all laughable because the scientists' ignorance is at such a higher level than ours that we are barely speaking the same language.

Quantum theory is ideal for this dubious purpose because ambiguity is at its core. In laymen's terms it suggests that reality--the essential, subatomic, invisible stuff of things--is either a wave or a particle depending on how it's perceived or measured. As anyone can see, this is wonderfully suggestive and I felt it freshly, almost as if it were a newborn idea, reading a passage in physicist Sean Carroll's *The Big Picture*. "The fundamental feature of quantum mechanics is that what we see when we look at something is different from how we describe the thing when we're not looking at it." Excerpted and isolated like this, Carroll's description can sound like a tease, but he goes on to nail it down: "When we measure the energy of an electron orbiting a nucleus, we get a definite answer, and that answer is one of a specific number of allowed outcomes; but when we're not looking at it, the state of the electron is generally a superposition of all those possible outcomes." The reason this statement was so arresting to me--although I've read it many times before in different language--was that as the words passed through my mind they conjured up a description by Malcolm Lowry of his difficulty in writing about an experience because as soon as he described it the experience was transformed and what he was writing was not what he had felt when he passed through it; in the process of recollection and writing it had become something else, something distorted or altered by his trying to express it. To correct this he'd have to start over, describe it again including this ambivalent feeling, only now he'd be describing something else because it had been altered by the first attempt. It meant that no matter how many times he tried to describe a single event it would always elude his effort by transforming itself anew in the process ("a superposition of all the outcomes"?). This notion that there is a field which represents life as we live it and that when we try to isolate any sole instance to study it, the nature of life--the field--is transformed, this notion is very appealing because it feels close to something we can feel. Life lived unconsciously is quite different from any self-conscious moment, even if we are outwardly doing the same thing. Or thinking or feeling or experiencing the same thing. Time can flow, taking us with it, or stop--pause I should say--while we ignore all else in the field of existence to focus on a sole perception or sensation. But none of this can be simply expressed because the moment the pen touches paper the experience is transformed. What you are trying to express has changed.

Someone who is not a writer can dismiss Lowry's complaint about his frustration in trying to render reality as a problem inherent in words, but for Lowry it wasn't a problem he could ignore. It expressed a specific condition that could overwhelm him when he was trying to capture experience in written form. The problem wasn't the words themselves but what happened when you stopped time--held a living being in place long enough to describe him and what he thought, felt, experienced--because to do that meant the original perception you began with had now changed so that whatever you wrote was false, a distortion of what you attempted to portray. Yet there was no other way to do it. The distortion or falsification or alteration was inherent. For Lowry this conundrum could be paralyzing. It felt as if you couldn't describe life without turning it into something it isn't. This has become a common observation. More recently cognitive scientists have learned that recalling a memory--telling it to yourself or someone else--can alter it. They don't know the specific chemical or electrical changes in the brain cells yet, but it seems there is an organic change when we make the mental effort to recall something. And that when the event or experience is recalled yet again, something about it seems different. Perhaps what's different is that there was an organic change in the neurons caused by the act of remembering, and a second recollection now includes the experience of the first one, besides the original memory. So these layers add up and may change the content of the memory too. Or embed it more firmly in the neurons (if that's where it is). So there is no one and only true memory--as vividly as one might feel it--and if we wish to preserve our own past by recalling it, documenting it, we have to accept some of the distortion that bothered Lowry. It's the best we can do.

26. Quantum reality

Adam Becker's book *What Is Real?* is about physicists grappling with just that question: what is real? It's no small mystery in the quantum era, even if it's about something no one can see, not even with the most powerful microscope. Quantum effects can only be seen indirectly as streaks of particles or energy recorded on a computer and reproduced on a screen. Physicists strain to understand what's going on in this invisible realm where the behavior of quanta--electrons in atoms--is so different from the "real" world around us.

I can't really understand the physics of what Becker is describing since I don't speak the language of quantum mechanics, and I can't switch to math to read the arguments in that alternative language, because I don't understand mathematics at any significant level either. So I'm stuck with the approximations in ordinary English, and as I type these words I'm aware of the awkwardness of my situation, using a symbolic language of hieroglyphs to try to express the physical reality of this baffling realm of the infinitesimal. Maybe this is just my problem--the way Lowry had his--but according to Becker physics has a very big one of its own.

The ancient question his book tackles--what is real?--arose in this new form when quantum theory was first being developed almost a century ago. But it was never answered; it was merely finessed. Left unexplained was the fact that the conditions at the subatomic level seem to be different from--in conflict with--conditions described by classical, Newtonian physics. The answer to this anomaly, according to the "Copenhagen interpretation," promulgated by Niels Bohr and his associates in the 1920s, is that everyday life operates under different physical laws than the quantum realm. In that invisible place reality is no longer relevant, said Bohr. This sounds like a strange argument for a physicist to be making since supposedly physics is the hardest of the hard sciences, the most firmly rooted in the real, measurable world--physics is physical--but Bohr's argument insisted this was something so different it could not even be described as "reality." This quite incredible claim was endorsed by the reigning math genius of his time, John von Neumann. He said his calculations showed that quantum theory was right, and as further proof quantum mechanics has been used ever since to solve some of the most critical calculations essential to our increasingly technical culture.

So the question about the fundamental nature of the real world has been ignored and even denied because of the theory's stunning accuracy in myriad applications, to use the kind of breathless language the physicists themselves use to pay homage to it. Yet mathematics is not reality, and von Neumann's own math was eventually found to have a critical flaw in it. The deep conundrum at the core of quantum knowledge has been slouching its way back to embarrass the physicists who still cannot answer this question: If you start at the classical level, in the world around us where Newton's laws rule, then move all the way down to the subatomic depths where quantum mechanics is the law, where exactly on this scale do the rules change? Where and how does classical become quantum? Does physical reality go through some kind of phase transition like water freezing into ice? Or are they parallel realities, and if they are does reality itself exist in more than one realm? That's an awfully big leap.

One of the remarkable aspects of the quantum mystery is that, as Becker describes, two of the greatest minds of the twentieth century squared off on opposite sides of the issue. Niels Bohr said we can't know what's going on with these infinitesimal particle/waves because their behavior is inherently unknowable, and can be predicted only as probabilities, not certainties. Bohr said these particle/waves of energy only seem to exist as we understand the word when we look at their traces. That's because it's a different world at quantum depth where there is no simple, certain "reality." In fact, said Bohr, the quantum world is not even real. Albert Einstein said this was absurd. He accused Bohr of copping out. The subatomic world is strange, yes, Einstein agreed, but it's part of the same one we're in, and one way or another it plays by the same rules as we do. We just don't know what they are. There is only one physics and what you're really saying by saying we can't know the answer is that the theory is incomplete. Something is missing, something we don't know, something that would explain this mystery.

Amazingly, even Einstein with all his prestige--he'd already won a Nobel Prize for his own work on this very same quantum theory--could not prevent Bohr from rolling over his protests using his own prestige and his poetic, almost dreamy portrayal of a quantum world where the usual questions about what was real could not be asked. Under ordinary circumstances it would be hard to believe that physicists would swallow such an airy argument but there is a part of all of us susceptible to the charms of magical realism and Bohr was, says Becker, charismatic. More persuasively, it has to be emphasized that in the background of this contest between two of the heavyweight physicists of the century, stood the authority of quantum theory in mathematical form with its ability to produce reliably accurate results better than any other tool. So Einstein's defense of reality was defeated.

That was a long time ago and although quantum physics has continued to be the gold standard for making many abstruse and difficult calculations, the anomaly at its heart has remained. And it has finally begun to embarrass the physics community. They would like it to vanish. As if this weren't frustrating enough, around us at their campfires are the heralds of the new Dark Ages, the priests and politicians and media personalities who feel threatened by those who unravel mysteries instead of bowing down to them. So doubt about what's going on in the heart of physics, say some of its sages, are best discussed quietly, out of earshot of those suspicious of science.

Since the issue is still unresolved, Becker is unable to end his book with a triumph. Yet he does show there have been theories created by inspired scientists that fill in the blank that Einstein insisted was there. They are flawed, or incomplete in some ways themselves, but they point a direction in the dark. Most tellingly, an Irish physicist named John Stewart Bell devised an experiment that confirmed unequivocally that there was an unexplained anomaly at the core of quantum theory. Bell's Theorem was a victory for Einstein's insistence that we must respect the demands of reality. There may be aspects of the quantum world that to us are "unreal"--the behavior of electrons inside atoms cannot be exactly compared to an apple falling from a tree--but they are part of the same continuous physical world. In that sense, they are equally real.

27. The most profound discovery

I'm trying to see if I can plumb Bell's Theorem. I'm going to this trouble, although it will profit me absolutely nothing in this world, because Adam Becker described John Bell's theorem as "the most profound discovery in science." This may sound too exuberant, although Becker is a physicist himself, but how can I tell if I don't even know what this profound thing is that Bell discovered? Becker's description of it, as good a writer as he is, failed to bring a bright enough light into my mathematically challenged mind, so I'm resorting to studying a website with some simple animation. Maybe pictures will help. My hope is that if I learn what Bell's Theorem is some window will open letting in a gust of highly oxygenated air. I know it's not likely because usually the explanation to this kind of mystery takes my mind a few steps in, then my concentration slips away vagrantly and by calling it back to attention I lose the slender thread of an insight I was barely holding on to. And now, what I realize after re-reading again Becker's description of Bell's discovery, and then looking at the animation on the website, is that what is interfering with my comprehension, besides my own mental limits, is expecting too much. I want a new dawn, a revelation, and I want it in rather literal terms. For the educated physicists there is a severe logic to the way John Bell's experiment is set up and how the results are interpreted. Becker describes what Bell's experiment proves about the strange behavior of quanta: "Bell's theorem really leaves only three unequivocal possibilities: either nature is nonlocal in some way, or we live in branching multiple worlds despite appearances to the contrary, or quantum physics gives incorrect predictions about certain experimental setups." As a consequence of this conclusion--and the last one just acknowledges that Bell's experiment may be wrong, so the choice is between the first two--says Becker, "Bell's work presents a threat to the Copenhagen interpretation." Nonlocal, I should explain, means a physical event that happens in one location instantly influences an event that happens somewhere else, as if they were touching. This would violate the law saying nothing can move faster than the speed of light. If they aren't touching yet affect each other instantly--how did the force bridge the distance between them? It means something moved but no time elapsed. Can't be. And the second choice requires that an infinite number of universes must exist. So Bell's Theorem shows that quantum mechanics as described by Bohr is either impossible or incomplete. His experiment has been done many times since and not been contradicted.

As I described before, the Copenhagen interpretation of quantum mechanics, as presented by Bohr and Werner Heisenberg, has been dogma in subatomic physics since the 1920s. Despite the weird dreamlike logic of twentieth century physics as expressed by Bohr, which says that there is no real quantum world and the "theory needs no interpretation, because the things the theory describes aren't truly real," Bell's contribution is now accepted as valid by mainstream physics. As Becker elaborates, "quantum physics proves that small objects simply do not exist in the same objectively real way as the objects in our everyday lives do...[so]...it is impossible to talk about reality in quantum physics. There is not, nor could there be, any story of the world that goes along with the theory." Bohr and Einstein seem to agree on the unreality of quantum physics but the crucial difference is Einstein's insistence that whatever is happening sub-atomically must follow physical laws even if we don't know what they are. (Maybe this is what he had in mind when he said, "God doesn't play dice." Things may look random because we don't know what physical laws they are obeying.)

As for my own attempt to understand Bell's Theorem, I watched the website video and appreciated its ingeniousness, and its clarity, which brought me closer, teasing me toward a revelation, but I'm still outside the club of those who can truly embrace its meaning. The practitioners have spent their lives thinking in the language of physics, both in words and mathematical symbols, teaching themselves to understand things that cannot be seen by human eyes but only, finally, glimpsed through sophisticated experiments, including entirely abstract thought experiments. It should be no surprise that compared to them I am earthbound. The surprise is the one Becker keeps coming back to, that the club of physicists has accepted the theory of quantum mechanics as whole when it isn't. Perhaps the finest minds can only be deceived by someone as smart and well-versed as themselves. That they can nevertheless behave like a herd, even all but destroy the career of an innovative and fearless colleague like David Bohm (whose work helped inspire Bell) adds a tragic dimension to the passage of quantum theory through the mind of man.

My own problem of understanding all this remains. I'm unable to understand Bell's great discovery intuitively, in a satisfying way, but I am at least pleased to learn that his findings are a victory for reality--for once I won't put Nabokov's quotation marks around it--and a faith in reality is the basis for my emotional investment in this whole controversy. Niels Bohr saying "there is no quantum world," so don't ask what it's like, offended my sense of the difference between what is real and what is not. I prefer Einstein saying there is a quantum world and it is a strange place but certain rules govern it, we just don't know what all of them are. So the theory remains incomplete. In pursuing my own limited understanding of Bell's Theorem, I've been reminded how much this distinction means to me. It may not be what Adam Becker meant when he said Bell's Theorem was the most profound discovery in science, but when I think of it as confirming reality by denying we can believe in a physical world which is not tethered to reality, I feel a deep resonance with what I assume is the truth about all existence. Even if we don't know the truth, the whole truth, about what it is that exists.

28. Salamanders

We caught salamanders in the shallow ravine not far from the main building at Pathfinder Lodge. Only Lee and I did this although other kids sometimes climbed up the wet rocks of the ravine. If you went all the way to the top it ended anticlimactically in some woods which eventually led back to the camp, so we stayed lower down where the stream kept the rocks wet. Most of the salamanders were orange although sometimes we'd catch a dark green one and wonder if they could change color like a chameleon. What fascinated me most about them was their delicate feet. The impossibly slender toes tickled the palm of your hand when you held one captive and their saurian heads were tiny miniatures of their Jurassic ancestors. We never did anything with them but catch them in the wet rocks where the stream trickled down the ravine. Once you caught one it didn't struggle in an end-of-the-world fight for freedom. They seemed to have a fatalism, perhaps because of their size, perhaps they knew they were so vulnerable they needed to trust we'd be gentle with them. After holding one for several seconds, examining the uplifted head, the staring eye, the impossibly tiny feet splayed out against the yielding skin of my hand, I'd open my fingers and lower it. At some critical point the salamander would sense the nearness of freedom, jump and be gone.

Lee and I were talking to one of the older counselors about the salamanders--excitedly, as if they were our discovery--and he told us they weren't salamanders, they were newts. For some reason this irritated me. Partly it was his authoritative voice, as if he knew quite a lot about lizards, and it was also his blandness, as if our discovery was no big deal. Most of all, the reason I felt irritated was silly. I liked the name "salamander" but did not like the word "newt." It was ugly and laughable, a nickname, as if the wee creatures weren't to be taken seriously. I thought the salamanders were beautiful and when they became newts they lost their beauty and gracefulness. The long musical name had been a kind of protection against its tininess and vulnerability.

Lee and I usually went to catch salamanders in the late afternoon when we had some free time with no scheduled activities. The path that led from the main lodge past the ravine continued on to an outdoor chapel where we had vespers every night. After supper, as the sun dropped, the kids began moving along the path to the chapel. The ravine was obscured in the shadows and walking past it I barely glanced in its direction. If I looked someone might follow my eyes and learn it was there. Lee and I had to protect the salamanders from discovery. That's what they were still called when I thought of them. The other name was only for when I was talking to the counselor who'd corrected us.

Pathfinder Lodge was run by the Baptist Church and Lee's father was the director. He often led the brief vespers service in the outdoor chapel. They kept it brief, I suppose, because of the mosquitoes and also because of our short attention span. A thought for the day, an announcement if necessary, a prayer, then a hymn accompanied by the most interesting feature of the chapel--the chimes struck with a mallet ringing out over our dutiful voices. The girls always sang with more commitment and gusto. The boys were vaguely embarrassed as if this was not their kind of activity. But the chimes added a special element. They were made of brake drums of different sizes mounted on a frame of metal pipes, the kind plumbers use. The man who made the chimes played them too. The still lake was visible beyond the chimes and a narrow span of beach. Lee's beatifically smiling father led us singing "We are climbing Jacob's ladder," and with its rising notes we began filing out of the chapel on our way back to the cabins we slept in. We were hardly "soldiers of the cross," as we sang our way along the path, and most of us would grow away from the church as we got older, but there was something about the homemade chimes making such a celestial sound, the unexpectedness of used brake drums being elevated to this sublime afterlife, manifesting transformation in this pure, humble way, that lingered with us as we walked back to the cabins.

It wasn't the sublime but the diabolical that lingered painfully in my mind after camp. The first day settling into our cabins, one of the other kids told me the boy assigned to sleep in the lower of the bunk beds where I'd thrown my duffel bag had just lost his father. "Lost?" "He died," the boy said. It didn't seem real. How could the boy be here at camp if he'd just "lost" his father? It wasn't possible. I moved my bag to the upper bunk and went out to look around. I didn't know any of the other campers except Lee and he was in another cabin. It was a warm summer day, ideal weather, and when I came back to the cabin there was a kid by the bunk where I'd put my stuff. He was a little taller than average with crewcut blond hair. He bent over his own bag of clothes and I said, pointing to the upper bunk, "That's my stuff there." He turned toward me, not saying anything, and I said, "You're the kid who doesn't have a father." His face changed and he swung his fist. It hit me on the cheek. Stunned by the punch, and stunned even more by what I'd said, I scrambled up in the upper bunk. There were some other kids in the cabin but no counselors. One of the other boys spoke briefly to the kid who'd hit me, then turned away and the kid went back to unpacking his bag.

I lay on the bunk, eyes damp, cheek hurting, but there was no blood. His punch was a gesture as much as a blow. What numbed the pain was my shame at saying something so blunt and thoughtless. I was also trying to think of a reason why he was in the wrong but as I recalled the scene I could hear the overtone to what I'd said: I have a father but you don't.

The more time that passed, the more shameful what I'd said sounded. What made it continue to hurt was the mystery. Since there was no premeditation in what I'd said I could only wonder where it had come from. I didn't understand it and couldn't explain it away. I had something diabolical inside me, something I couldn't control and wished wasn't there.

29. Mutual Assured Destruction

William Perry's autobiography, *My Life on the Nuclear Brink*, took me back to 1962 instantly. That was the year of the Cuban missile crisis, when we came very close to setting off a nuclear war. It was the year after I graduated from high school and "Trapped on Earth" might have been the name of my account of that year. Where could you go to escape the threat of thermonuclear bombs? Nowhere. Forget it. We're all stuck here.

I was also aware--I think I was aware--that my preoccupation with the threat of nuclear war was partly a pose. The disparity between what a nuclear war might do to millions on the planet and what it might do to one individual shows the shameless self-regard of a teenager who sees nothing odd about taking the H-bomb personally. (It also recalls Stalin's famous quip: the death of a single person is a tragedy, the death of millions is a statistic.") My pose, no matter how deeply rooted in the truth about the human condition, strained the patience of adults, no doubt. What was anyone going to do? (New York's Governor Rockefeller wanted us to build bomb shelters in our backyards.) Did William Perry's preoccupation with the threat of nuclear war strain anyone's patience? I don't think so. Unlike me, he was doing something about it. Working to prevent it became the goal of his career.

He seems to have been working to make himself unnecessary. Only a man with a fully developed sense of himself and a deep natural modesty could dedicate himself to such a self-effacing ambition. And Perry was an ambitious man--he became Secretary of Defense--but he seems to have been an unusually mature and mentally healthy man. He was not seduced by the sound of applause like his boss, Bill Clinton. Like most presidents and many politicians, Clinton's sense of himself was incomplete or inadequate; it needed feeding and filling. Or perhaps it was voracious. Whatever the reason it needed the approval of the crowd to be satisfied. This must be the same need any performer has. As Obama said, all politicians have egos. It's an unanswered question whether a man like Perry could serve as president. It's unlikely that experiment will ever be conducted since the political process is so unlikely ever to produce a man like him for the office. At the other end of the scale, where the need for applause nearly consumes the whole person, many candidates have appeared, draped in their need, most notably that egregiously lower case man, donald trump. Perry is his antipodal opposite. Which means he isn't good theater. He does not stir our emotions. That fact says much about the limits of democracy.

I wouldn't have read a book like Perry's memoir back when I was most afflicted by my nuclear headache. It was too boring. This is a compliment. On this extreme subject, nothing else is appropriate. There is no soundtrack to Perry's sober book, no music plays as you read. From this vantage, all these years later, you can see how successful Perry and all those who worked with him were in preventing us from blowing ourselves up. If there is a hero of the era Perry is describing, one besides himself, it would be the Russian colonel who decided not to launch nuclear bomb-carrying missiles when he saw his radar screen picking up signs of what looked like a nuclear attack on his country. It was an act of courage made under extreme pressure. Although it turned out that the radar he was staring at had picked up a stray artifact, not a lethal missile, the colonel could have been wrong. Maybe a kind of fatalism guided him. Maybe he thought, "It could be a surprise attack by the Americans, but there is no current crisis between us and if I'm wrong I probably won't be alive to regret it. I don't want to be the one to set off World War Three." The threats of nuclear war are so unique, so unlike anything human beings have had to adapt to, they often lead to a fatalistic attitude. They expose our delusion that we are in control of our own destiny. Yet with the next breath we remind ourselves that it is unhealthy to believe we can do nothing, even if it is mostly true, because doing nothing is giving up, and who wants to give up? The two poles are voiced by

William Faulkner and Samuel Beckett. Perry quotes Faulkner's famous Nobel Prize speech, "There are no longer problems of the spirit. There is only the question: When will I be blown up?" Faulkner answers his own question by saying: "I decline to accept the end of man....I believe that man will not merely endure, he will prevail." And Beckett, whom Perry does not quote, crying out from a landscape that might have already been bombed, gives his grudging salute to the human spirit at the end of *The Unnameable*: "I can't go on, I must go on. I'll go on." Beckett's voice is not giving up, not shutting itself off, although he has no hope and ends with a sigh of resignation, far from Faulkner's brave defiance. Yet neither writer says why he is going on. One goes on because he must, the other because he refuses to quit. The life force is speaking through them in two different voices, although I've drained the poetry out of them both. Draining the poetry is my goal, to see what is left in the dregs, to see what we have in common with the voiceless, mindless pullulating unicellular life we all came from, only to end up where we are today, self-conscious, sentient beings contemplating a global immolation called *mutual assured destruction*. Which might be another name for the sixth extinction of life on earth which some scientists see as the next looming self-sacrifice on our civilization's horizon.

30. Invisible

A verse from I John 4:12 was quoted in the newspaper and it jumped out at me for its modernity: "No man has ever seen God; if we love one another, God abides in us and his love is perfected in us." We can easily read this to be saying

God is a concept, one that is consonant with the notion that man created God, our greatest conceptual creation one might say (along with money). A footnote in the Oxford Annotated Bible supports this contemporary spirit, saying, "To love one another" is the only authentication that we know God, "whom no man has ever seen." In other words--in words of compassion instead of metaphysics--there is no God unless we create him by loving our fellow man and this love is the only proof of his existence. As they used to teach us in Sunday school, "God is love."

This is a subtler, more human as well as a more practical worldly "proof" for the existence of God than Thomas Aquinas' handful of reasoned arguments because it doesn't try to be logical. It doesn't invite refutation or disputation. It is a simple, irreducible description. It may be wishful thinking, or sentimental on a cosmic scale, but it lives in the world of actual behavior, not theological argument. It is the message Beth and I carried out of church with us. We couldn't carry it very far into the rest of our lives for many and varied reasons, among them that the pure simple message wasn't so pure or simple when you lifted your eyes from the page and surveyed the human landscape. Or if your eye wandered over other nearby passages from the same chapter, you discovered the context of the verse entangled it in the coils of superstition we were trying to escape from. "The reason the Son of God appeared was to destroy the works of the devil," says I John 3:8 just a few verses before, and assertions like this surround the one I quoted so that in context it seems blotted out by the shadow of primitive mythology, the devil and sin and the dark, fallen world Jesus was sent to save us from, according to the myth of resurrection. Scripture like this can be heard in the rants of the evangelical preachers today whose minds seem blind to even the faintest rays from the Enlightenment and everything we have learned in the past three hundred years of spectacular growth in scientific knowledge. Nevertheless, a quotation like the one in I John 4:12 can be a bridge from the past to the future that makes us feel there are underlying constants in human experience, and perhaps they will even be relevant at times in our gene-engineered future, these hints that there is something to nurture, even cherish, in our strivings on planet earth. Nevertheless, I don't entirely believe it myself, which is why these few last words don't sound as if I wrote them. They could be quoted out of context, just as I quoted the passage from I John 4:12. Viewed from a pyramid in space, there is something touching about how we poor lambs on earth can imagine that a supernatural force takes any notice of us at all. The day after I read the quotation from I John, reminding me that if we love one another, this particular God exists--"abides in us"--I read an account of a Muslim making the hajj, the pilgrimage to Mecca. The Muslim pilgrim was Diaa Hadid, a reporter for the New York Times. Her account, as both a believer and a writer for a secular newspaper, was respectful reporting with more private details than usual in a news story. For instance she told us what she prayed for when in the inner sanctum of the Kaaba, the Grand Mosque where the black stone once shone whiter than milk, but now, stained with the sins of man, it rests darkly on its pedestal, awaiting the kisses of the faithful. They are assured their prayers in this sacred place have one hundred thousand times more power than prayers offered from a less sanctified place. Juxtaposed with this very traditional, even primitive faith in the power of place, is the Islamic insistence that because God has no face, no body, no corporeal form, man can make no image of him without committing a sacrilege. So the holiest place in all Islam is empty. The reporter says to her this is a powerful idea. It's one that speaks to me too, perhaps in another voice from the one Diaa Hadid hears. Or perhaps not. It may be that both of us respond to the intuition that what is most exalted must be invisible by its very nature. Yet the human element in her pilgrimage can't be ignored. The visit to the holiest shrine ends with a treat of ice cream and outside the mosque "children hawk velvet prayer rugs decorated with images of the Kaaba." The need for an image of some kind as a souvenir for the faithful is testimony to our need for the visual, the actual, for a palpable totem, and I don't think it detracts from the abstract quality at the heart of Islam although it seems to flout the strict rule. In a part of the world where nature is so spare and seems to embody opposites--it's either hot or cold (mostly hot), it rains or it doesn't (mostly doesn't), the land is sand except for rivers or oases or stone, an extreme binary topography in which the world is all one thing or the other, and you are saved or damned with no state in between.

If the modern mind has left religion behind so our description of any metaphysical existence is in the language of particle physics and cosmology, the place the two belief systems intersect may be their agreement that ultimate reality--the "truth"--is invisible, beyond detection by the human senses. We believe the descriptions of reality according to cosmology and quantum physics (despite their incompleteness) because their priests have discovered things we could never have learned without them, and these things have transformed our world. Most of what they have learned we must take on faith because we can barely understand what the sages are telling us they have discovered. We share a portion of the terrific curiosity which drives them on, yet the forces they feel so compelled to know are utterly unaffected by our unquenchable interest in them.

31. Ignorance

"You hired a blind director?"

The executive producer was shouting into the phone, into my ear.

"Well, yes, but no, not really. I just said that to get your attention," I laughed. It had been too tempting to pass up.

Russ didn't hear what I was saying anyway. He enjoyed shouting. It was recreational. There was another guy he worked with who enjoyed shouting too and sometimes the two of them would spend all morning pummeling each other with arguments, break for lunch, and come back in the afternoon for Round Two.

As I explained to the executive producer when he quieted down, I only hired the blind director to supply a crew and equipment, not to direct. I was going to handle that myself. It was a simple shoot. There were a couple days of meetings and lectures followed by a big dinner with speeches, then they all went home. We'd pick up a couple interviews in between shooting selected lectures. It was a low budget show on one of the dullest subjects on earth, yet as it's sponsors kept repeating it was central to the development of the educated person and to the health of the Republic. It was called General Education or equally blandly, "the common learning," which didn't even sound like a name. It was like a multivitamin containing what the educated person should know. The educators had carved this subject into different slabs and the prevailing emotion at the colloquium, as they called their gathering with Latin dignity, was earnestness. Perhaps it was this church camp mood that inspired me to hire the sightless Howard to "direct." I was also getting back at Russ, in a backhanded way, for hiring me to produce this dull project.

The day before I had this offbeat conversation with him, I was standing in front of Hutchinson Hall on the time-warped Gothic campus of the University of Chicago waiting to meet Howard. We had only talked on the phone and he hadn't said anything yet about being disabled, so I didn't know who I was looking for. When I saw a man wearing a baseball cap and using a cane to walk slowly up the sidewalk, I had a disquieting feeling this was my man. Yes, the cane was white. Walking with him, holding his arm, was a young woman so perfectly beautiful she might have been on her way to be filmed for something herself. Something commercial, like an advertisement, not a colloquium on education. She was guiding him up the sidewalk toward me and gave a small wave of recognition, as if we must be the ones looking for each other. She was so pretty anyone she waved to was going to respond.

"Howard?" I called out. "Howard Friedman?"

"Hi, Eli," he said amiably, as if we were old friends who'd worked together before. "This is Brenda. She works with me. I've got this temporary problem with my vision," he went on, as if it had just developed that morning, like a cough or a limp. Brenda filled in the details later. Howard had serious allergies. A few days ago, he'd taken his usual drug along with something new, a steroid maybe--she didn't know--because his immune system occasionally got out of equilibrium. He lay down for a nap and the next thing he knew he was blind. Something about the optic nerve attacked by his own body. The doctor expected it to go away once he got the new drug completely out of his body and was restored to something like normal. He was allergic to all kinds of things and had unusual problems at times. Meanwhile, he carried on his business as well as he could, helped by friends and contacts in his business. He owned a television production company and had a lot of equipment one could rent.

Brenda didn't need to say that Howard obviously had some chutzpah and was not going to let a passing case of blindness prevent him from taking a job. He was also a good psychologist. After meeting him in person, how could I say no. Well, I could, but I wasn't going to. I was glad to let him make a few bucks from our production, renting his equipment and using his camera and lighting crew.

It didn't hurt that Howard had such an attractive and likeable woman helping him. It bespoke some invisible qualities in him despite his disability. As the three of us walked around Hutchinson Commons "scouting" the location,

Brenda unobtrusively warned Howard of surprises he might trip over. It was touching the way as we approached steps up or down, she'd quietly interrupt to say: "one step up," or "one step down" or "stairs going up." Near the top she'd say, "one more step." He followed her directions (the blind director!) and acted as if nothing was wrong. In the spirit of General Ed and earnestness I was pleased with the decision to hire Howard and his company. It was giving the shoot a dimension it had lacked before. The crew he provided was competent and friendly in what I thought of as the midwest style. People in Chicago were nicer than what I was used to back east.

We filmed during the two days of lectures and presentations, including some interviews. At the end of the third day we set up our cameras in the monumental dining room of Hutchinson Commons with its wood paneled walls hung with portraits of historic university luminaries. The star of the colloquium and featured speaker at the closing dinner was the scientist, physician, and writer, Lewis Thomas. Dr. Thomas was introduced by the university president who praised Thomas for his knowledge of science and the humanities, as well as the breadth of his other cultural and intellectual interests. He was a compleat scholar in the best tradition of general education. The audience of professors and educators was receptive, themselves another ideal of the human kind. Besides being comfortably sheltered in this great neo-Gothic hall, a sanctuary of civilization, they were well fed and mildly lubricated with cocktails and wine. The earnestness was flowing, or congealing, in conducive conditions for contemplating nourishment for the human soul. Lewis read his speech in a light, pleasant voice with a self-assured yet tentative timbre fit for his subject. He knew where the laughs were and how to lead up to and fulfill a narrative payoff at the end. He made you think if everyone had a mind like his, the whole world could be a sublimely civilized place for us to live in, reasonable, intelligent, inspired, magnanimous.

I listened from the back of the great hall where I could see the three cameras and their operators, but they were recording on their own and we'd sort it all out later in the editing room. With nothing to do but watch and listen, I gave myself to the speech. It was such an artful thing of beauty that, besides the laughter, once or twice I had tears in my eyes in reaction to Dr. Thomas's wit and poetry. Starting with his suggestions about how to teach science and produce doctors, Thomas moved on through an assessment of the major questions of today--how we know and what we know. It was a paean to the constant and creative yet unsettling churn of science:

"Order in the universe at large is preserved along with order at the innermost benthic depths of living cells, and the infinitely small working parts of the atom, no matter how many firmly embedded ideas about how the play works are destroyed by the new bits of information brought in by science. It is a coherent world. We know this in our bones, although we continue to find in it a complexity beyond our comprehension. It fits and holds together and this is the central dogma, and every new scientific observation contributes, sooner or later, to the stability and the imperturbability of this underlying truth. The ways in which it works, the details, are always a surprise but only transiently, and at bottom is the solid idea of order and coordination taken for granted and assumed."

This was reassuring, but interrogating nature was not a complacent activity, no matter how earnest:

"The greatest of all the accomplishments in twentieth century science has been the discovery of human ignorance. We live as never before in puzzlement about nature, about the universe, about ourselves most of all. It is a new experience for the species. A century ago, after the turbulence caused by Darwin and Wallace had subsided, and the central idea of natural selection had been grasped and accepted, we thought we knew everything essential about evolution. In the eighteenth century there were no huge puzzles; human reason was all you needed for figuring out the universe. And for most of the earlier centuries, the Church provided both the questions and the answers neatly packaged. Now, for the first time in history we are catching glimpses of our incomprehension. We can still make up stories to explain the world as we always have, but now the stories have to be confirmed by experiment and then, once confirmed, reconfirmed. This is the scientific method and having started on this line, there can be no turning back. We are obliged to grow up in skepticism, requiring proofs for every assertion about nature and there is no way out except to move ahead and plug away, hoping for comprehension for the future, but living in a condition of intellectual instability for the long time being."

The pleasure of listening to Thomas's oral poetry would not, it occurred to me, have been shared by Howard, the blind director, had he been there to hear it. To lose your sight to the unexpected interaction of two products of the scientific method coursing through your body would have made the poetry sound ironic, if not downright perverse. Instead of enjoying a miracle of medical science, Howard was a victim of an inadvertent experiment which had, at this point, tragic consequences. If he had been sitting next to me in the great hall he might have had tears in his eyes too, thinking that when the miraculous goes wrong, the power of science could be as cruel as anything nature dished out by chance. However, Howard wasn't there. I was free to savor the pleasure of Dr. Thomas's eloquence. It was a feeling akin to what I'd felt at times, although rarely, so many years ago, listening to a sermon. It made me feel connected to something larger than myself. Now all that was unrecoverable history, lost in the jostle of experience, but making the connection helped explain why I had been so willing to hire the blind director, and why doing it had given me so much satisfaction.

32. Odds against the universe

In John Boslough's book on Stephen Hawking's cosmology, there are repeated quotations and paraphrases of Hawking saying how improbable the universe is. There are about fifteen numbers, all with critical values that have to be within a certain narrow range and if they weren't, even just one of them, the universe either wouldn't exist at all, or would never have produced sentient life like us. The odds are, to use the apt word without exaggeration, astronomical. (Martin Rees says there are just six of these values which were imprinted in the big bang and if any one of them were "untuned" there could be no stars and no life.) These narrow conditions might be why Hawking uses the word God at times to mean the force or entity or something-or-other that operates behind the last veil we can see, measure, trace, or intuit. As a scientist he has no word for whatever it was that finely tuned all these physical values so existence could exist, balanced so keenly between expansion and contraction for these billions of years. But does he really think there is some supernatural force or influence out there (or in here) which can only operate indirectly, through these physical forces with such power yet so strictly limited? The problem with asking a question like this is that it takes us beyond language. By that I mean the words are asking something that can't be answered in words. In that sense, the words deceive us. Not with an answer but with a question.

You get a hint of this conundrum when you read Boslough's description of the "anthropic principal." Because a description of physical reality before the universe existed (that is, an answer to the question, "What would be there if there were no universe?") is not possible. Hawking says instead, "If there were no universe, we wouldn't be here to see it." To see its absence. To not see it. (This is not as silly as some other statements of the anthropic principal.) Since we are here, there is a universe. That isn't as satisfying to say. The bones of the tautology begin to show through, as they always do in this game. Or, if you prefer you can simply say, the nonsense is evident. Yet it has a logic. The physicists say that in the last instant before the universe contracts and does what we don't know, all the forces merge into one, and time and space ("space/time") vanishes. The magician disappears in his hat, taking his hat with him.

Even if it's frustrating, it makes perfect sense to me that there would be a point beyond which our understanding could not go. Hawking's goes much further than mine but it too has a limit. At the limit it boggles. If there is something supernatural beyond the limit of nature, beyond the limit of our understanding as products of nature ourselves, then we don't know what it is, what form it might take, if it has agency, force, power--we don't know anything about it except its mystery. Could it be an intelligence which is using the universe to create life so it can...do what? Fill in the blank. Anything goes. Except certainty. We can say nothing about it for certain. We know nothing of any intentions besides our own. There is that wall of chaos--no, not a wall, a state or passage between the universe (the physical) and whatever precedes it, if anything. We can't assume it intends anything. Only life exhibits intention. Which may be a force that emerged from life after it happened. Existence may have produced the urge to exist. It may also contain the urge not to exist. Maybe.

My description is inadequate; I can't say it satisfies me. I can't say anyone's does. I find I can keep revising the words endlessly, until any meaning seems to have been spun and squeezed out of them.

33. Dad's wall

On one of Dad's trips to the hospital earlier in his long slow motion decline, they had had to tie him down to the bed. Otherwise, he'd get up and go to the bathroom by himself, or walk down the hall (he didn't need a walker then). Their fear was he'd fall and hurt himself. He didn't wander, he didn't have Alzheimer's, he had vascular dementia. He was restless and just wanted to walk around, to be free. "I don't fall!" he'd say in a louder voice than we'd ever heard before. This fierce person was emerging from the stripping away of normal abilities. He had perpetual scabs and healing places on the thin skin of his forearms where he'd bounced off things, including walls and sometimes the floor, after losing his balance. The self-sufficient man who took care of himself and never needed help, never asked for it except in the phrase, "Say, Eli, I could use a hand this afternoon," commanding my labor as much as requesting it; that man was disappearing. His absence had left Dad a vestige of himself and despite his poor memory it was apparent an inkling of his former whole self was still there, if barely. This faint trace of former competence gave him no comfort. It was a diabolical irritation, like the memory of what it was like to drive a car.

The ferocity and single-mindedness of his residual self reminded me of when he built the wall at the cottage. There was something biblical in the way he hurled himself against the challenge of building a wall out of railroad ties to contain the rocks at the front of the beach. The cottage Mom and Dad bought was on one of the Finger Lakes and the storms weren't oceanic, but they usually came from the north end of the lake and pounded the beachfront head on because of the angle of the shoreline. Dad had a truckload of used railroad ties dumped at the bottom of the ramp at the end of the beach and using an old pair of ice block tongs he'd grab an end, drag the tie into the water, then float and heave it into place. He was not a big man and only had me to help him. I was a querulous teenager who tried to avoid him, although the times I helped I usually got into the spirit of his ambitious, self-imposed labor. He was patient, steady, analytical, he didn't curse, and always took the more onerous part of the job for himself. I couldn't help but admire his virtues. The next door neighbor, a vigorous outdoor guy just a few years older than me, volunteered to help more than once but Dad just said, "Thanks, I think I'm okay." It was his project.

It took most of the weekends during the summer and a couple weeks of his vacation to finish. The ties were secured in place by steel pipes Dad drove into the lake bottom with a sledge hammer. The pipes passed through holes in the ties he drilled by hand with an old brace and bit, and the ties were stacked on top of each other to make the wall. He also put in several ties at right angles to the wall, extending back toward the cottage, with more pipes holding them in place. The whole construction was about sixty feet long, five ties high at its peak. It wasn't deep enough to dive in from the top of the wall, except maybe at high water in early spring when it was too cold to swim anyway. There was a dock neatly jutting out into the lake from the middle of the wall, and from the end of the dock you could easily dive into the water. The finished wall was a huge success, very close to what Dad had envisioned.

Each year in the spring everyone along the lake had a story about what winter had done to their beachfront, but Dad's wall survived very well. He made repairs over the years, especially after an unusually cold winter or with higher water than average. The expansion and contraction of freezing and thawing, combined with the battering force of storms rolling down the lake did its damage, and over the years the wall got lower and shorter. He repaired what he could and was often able to fix dislodged ties until the last year they owned the cottage, nearly fifty years, when a record storm left a pile of ties looking like pick-up-sticks on the beach. The storm was a signal their time was up as owners. Looking at the mess on the beach, the debris of what Dad had built, the order he'd created in single combat against the indifferent and chaotic forces of nature, I felt a swell of pride in his stamina and stubbornness. There was no hubris in his task. He didn't overreach, or if he did not by much. Instead of dominating, his building projects had more of a Zen-like quality. Whatever he built was always neat and economical; he tried to work with what was there and not expect too much. You were not going to defy water and ice and the laws of physics. However well-constructed, it would have to be repaired at times, or rebuilt, sometime. His wall lasted almost half a century, all the time they owned the cottage. That was enough. When he looked at the remnant left on the beach after the final spring, he said, "It's someone else's turn now."

34. Slides

The year after Dad retired for the second time, this time from teaching organic chemistry to aspiring young scientists, doctors and nurses, we got them a new slide projector and screen for Christmas. During all those years when they were raising us, Dad took pictures, mostly 35mm slides, of all the family activities. I was quite alienated from the camera for the private reason that I didn't want to be part of this family and culture. I felt guilty about having this feeling but couldn't help it. As soon as I saw the camera in Dad's hand I frowned and "made myself scarce," to use the kind of figure of speech Dad found amusing. He wasn't amused by my subversive attitude, but it could be ignored because everyone else was quite content to be documented being themselves on birthdays, holidays, and most any occasion. The cell phone camera has replaced 35mm slides nowadays and the documentation of one's own life begins even earlier today than it did sixty years ago. The new technology makes it cheaper, or rather it feels cheaper because you've already invested in the phone, and it is even handier than a reflex camera. You can document everything. The photos accumulate and you save them digitally in a personal history of our endlessly self-regarding world. One day, you think, I'll want all these records of who I was and what my life looked like. And maybe you will. But here is how nature surprises us. Dad was always most cheerful shooting slides of kids dressed for Halloween or blowing out candles on a birthday cake or water skiing or snow skiing or with dogs or cats or hamsters or in Scout uniforms or grinning to show off missing front teeth. Yet when we set up the projector and screen in their living room in Florida and opened the first metal box of slides and started projecting scenes of a typical day at the cottage, Dad was quickly so overcome with emotion he was in tears. It wasn't because his eyesight was so bad he could barely see the images on the screen. That was true too, but it hardly mattered, and may only have reminded him even more what was missing. With barely a glimpse he knew just what each slide was showing, and the sweetness and poignance of memory we'd hoped to give him with a slide show of his own photos had turned into a picture of loss. Instead of happy recollection, the once-in-a-lifetime transitoriness was a cruel portrait of what was gone. Yes, he had had a good life, even at times a wonderful one, especially compared to his own father's, with five healthy kids all raised in comfortable homes full of innocent fun as we whisked through the days of growing up. Those days of unself-conscious coping with daily life were what he had cared about more than anything, and here it was on the screen right in front of him, gone, unrecoverable. He had loved it so much he couldn't stand to see it again, and the emotion he couldn't express because of dementia and lack of speech came out in tears.

While Dad wept, Mom sat on the other side of the room, narrating her own recollections: "Oh, I remember when we got you that bathing suit, Jeannie--it was just before we took you to camp where you got such a bad case of poison ivy. Remember?"

After a few more slides, we shut off the projector and didn't look at them again.

35. The right whale

The right whale got its name, Melville says, because it was the "right" whale to hunt, the right size, plentiful, killable, full of oil. After so many years of uncontrolled hunting by humans, the declining population of right whales has probably also been accelerated by the reduced volume of phytoplankton in the warming waters of their breeding ground in the Gulf of Maine. News stories are warning us they are heading toward extinction. Only 430 are left according to scientists, and this year no females produced pups. Zero. Those who study the biological world and produce these statistics know what this means. The article in the Boston Globe didn't say it, but it occurred to me that the whales may know what is coming too. They may be like the Marquesans Stevenson met who lost faith in their culture and whose bereft children quit learning the words to their traditional songs. Since the whales are social animals, they may sense in some way--and express in communication with each other--that this world is no longer encouraging them to be fruitful, multiply, and enjoy the boundless sea. Besides the diminished food supply and warming sea, the right whales are menaced by entanglement in fishing lines. The lines tied to lobster pots are especially dangerous. When the lines are wrapped around their massive bodies, the whales have to work harder to swim. This increases the stress on them and stressed females don't reproduce as plentifully. In other words, they get depressed by the fishing lines cutting into their skin and making swimming an ordeal instead of natural, even joyous. The changed conditions for life as a right whale have been measured over the years by scientists. Over many years. It has taken a long time for us to bring the once most abundant right whale to the gate of extinction. Instead of it being exceptional that the whales have gotten the message about their fate--and in a more immediate form than the scientists have measured it--I think it would be merely hopeful, even cowardly, to assume they don't know who is doing this to them.

36. Stochastic

We all know evolution is random. But maybe it's wrong to assume there is anything everyone can agree we all know, so I should just say evolution is random whether we know it or not, and out at the edge of the one abiding purpose for animate existence, to reproduce itself, there are instances of side effects where a trait not aiding survival has itself survived. Like how did we end up with such an excess of consciousness, with which I'm including the tendency to become spellbound in a reverie at odd, involuntary times. It isn't just the brain taking a rest, like sleep, I don't think. It seems to grow out of a moment between attention spans, a moment stolen from the serious business of staying alert to survive. It isn't active like play, or intense like concentration, nor is it a state of logical thinking. Instead of a linear form like a train of thought or a narrative, it seems shapeless, amorphous, a waking dream oozing along by associations you couldn't make up--and if you try to control them the spell snaps and you are dropped gently back into ordinary consciousness. Since reveries are pleasant, we don't take them very seriously. They are only a time out for the brain before returning to the business of paying attention. Having discovered this role for the pyramids in space--to serve as a reverie inducing perch--I think it devalues this mild, harmless virtue to try to inflate them into something they're not. Especially since one of the things they are not is full of hot air. A reverie itself can be a great deflator. A group of people sharing a state of reverie is, by some definition, a democracy in the same way a room of sleepers are all equals. The difference is that sleep is recognized as a necessity and a reverie is a luxury, or as I called it before, a stolen moment (a nod to Oliver Nelson), an AWOL mental blink from where your mind is supposed to be. It's like playing hooky, with benefits. A person lost in a reverie is not one who can commit a crime. He, or she, cannot do violence to another living thing. For the duration of the reverie, the person is more than absent, they are bodiless, abstract, as lost in space as that pyramid out there that may have inspired their state of mind. There is a quality so celestially harmless about the reverie that I think we should all pause for a moment and give it its due. That exalted blankness is sometimes identified as the inspiration, or the encouraging atmosphere for a fresh idea to appear in an ideally ready mind. This is not an accomplishment anyone can take credit for since it arrives on its own, unbidden and unbidable, and there is something quite inappropriate about insisting it must be useful and have a purpose. So, switching to the other side of the hot blue dot, maybe we should just say it's one of those Zen moments, a gift of some kind from one's tricky mind, and instead of rushing to explain it we should all just lie back, fall silent, and sigh.

37. The Wall of the Crow

The wall is a low enclosure around the major Egyptian pyramids and the Sphinx, also enclosing a few smaller buildings used for maintenance and administration at the site near Giza where these most famous and most visited relics of ancient Egypt are located. This enclosure appears to honor the crow, the distinctly uncommon common crow, the one I wrote about in a book some years ago. It's not a scientific book, there are few facts in its pages, although there is a story, a search for an answer to a question. I took this pursuit in a fictional direction because I don't have the scientific knowledge to investigate why crows have more brain power than they need to survive. It struck me as an interesting question, however, and can be asked about a number of species like chimps, bonobos, dolphins, orcas, ravens, parrots, and even octopuses (yes!). If you looked into it, the list would grow, not shrink, and keep growing the longer you studied animate life. It's more evidence of human arrogance. We're sure we're not only the smartest creatures on earth but also the only ones who can think.

The wall in Giza around the massive monuments to human vanity, ambition, and fear of death is the rebuilt remnant of a wall where three or four millennia ago the crows may have gathered as the pyramids were being built, watching the bipedal laborers and commenting to each other on all the curious, or merely puzzling, things they saw. Maybe the wall was such a humble construction compared to the grandiose structures it enclosed that the pharaoh didn't even bother to name it, and it became known by what the workers descriptively called it. In this way it might have been haphazardly named after the black feathered audience which had kept them company during the long years of construction, accepting the small handouts from the laborers' lunch baskets when they all took shelter from the hammering sun to eat a meal, then lay down in a shadow for a nap before going back to work on the mightiest tombs ever built.

What I also find curious about the Wall of the Crow is the possibility that the name is as ancient as the modest structure it refers to. Just as the Sphinx and Giza have names whose origins may be traced back almost three thousand years to whatever they were known by when they were built, so this modest enclosure may commemorate the rude, noisy, black-feathered witnesses to the original construction. The crows may even have their own memories, and they may be proud of their venerable ancestors who were there at the beginning. They may consider themselves experts on how far a human being will go to glorify himself, and may even flatter themselves that when it comes to the glory of the animal kingdom the ones with wings and black feathers have more wisdom than the ones who are so proud of their brains and their thumbs. The stones of the pyramid were the pharaoh's monument, but the eyes of the crows watching the laborers raise them up into the sky may have made the corvid consciousness part of the workers posterity. And, as if they were aware of this, they named the crows' perch after them.

38. The Svevo Fallacy

Stephen Hawking's *A Brief History of Time* says: "...the discovery that the speed of light appeared the same to every observer, no matter how he was moving, led to the theory of relativity--and in that one had to abandon the idea that there was a unique, absolute time. Instead, each observer would have his own measure of time as recorded by a clock that he carried: clocks carried by different observers would not necessarily agree. Thus time became a more personal concept, relative to the observer who measured it."

Yet, even if each of us keeps his own time, so to speak, the speed of light remains the same for all observers. What Einstein at first called the "invariance" theory is now commonly called the theory of relativity. It became known by the quality that applies to us, or so we think, not to the facts of its physics (which most of us don't understand anyway). We have spent the century plus ever since finding ways to apply "relative" or "relativity" to things that once might have represented what we assumed was stable and unchanging. In the new regime of assumptions about our cosmic home, nothing was stable and unchanging in the way we'd assumed and we learned we should remember that fact and apply it everywhere. It's the triumph of what we could call the Svevo Fallacy. Because what happened to the speed of light in our new formulation? Isn't it supposed to be the same for all observers? Isn't it the one thing we can "believe in" as constant?

Words stutter when used to describe the behavior of invisible physical forces, so there's always a question about whether we should take descriptions of "reality" from physics and apply them to our own experience. Flesh and blood may be made of atoms, but experience is something different, with no reified form, just as a mind is emergent and intangible while a brain is an actual thing. But Hawking's description was one I could apply with a particular spin, so I borrow it for a purely personal purpose. Now, I've forgotten what that purpose is, but the puzzle about whether "relativity" means something is relative or invariant is intriguing, so let it stay.

39. Many worldz

Adam Becker's book suddenly irritates me by taking seriously the notion that there are many universes besides our own surrounding us invisibly. He's been using this idea throughout the book. One of the theories he cites to explain quantum theory, Hugh Everett's universal wave function, requires the "many-worlds" condition. I sort of ignored it because I sympathized with the role Everett was playing in Becker's story, and he's the physicist who came up with this explanation. Then I abruptly reverted to my hostility to this idea when Becker began referring to it in a way that reminded me of the string theorists' blithe invocation of an infinity of universes and I recalled the host in the PBS series on quantum theory passing through a wall (a special effect, not a "real" shot). When sub-atomic physics violates biological integrity, I jump off their wagon. (What is biological integrity? It's the assumption that a living animal made of blood, bones, and guts can't walk through a solid object.)

Isn't this just a flat-earth reaction on my part? I don't know for certain that there aren't many universes (no one does) but a number of theoretical physicists are finding this idea implied by their calculations about the nature of matter and energy at levels way smaller than any human can see.

There are too many of these appearances to merely reject them. That's the flat-earth response. Besides that, these people are smarter than I am. Any physicist is smarter than me; it's a physical fact. It can be proved mathematically. There are also equally proficient physicists who are unconvinced by the "multiverse" proposition.

What is more interesting than whether the many worlds theory is true--it won't be decided in the shrinking tranche of what's left of my lifetime--is this reflexive rejection of ideas one doesn't like. Where does it come from? What does it say about us? Is it just a resentment of people who are smarter than we are? (By smarter I mean better at thinking with numbers.) It is probably related to our preoccupation with status. Someone smarter than me has a higher rank in the species, a superior right to resources, to telling me what to do.

I learned there were people smarter than me in 12th grade when I took Advanced Algebra. There were ten of us in the class, all males, and in a school that had plenty of smart kids this was the most concentrated group of brains I'd met. Kids I didn't even know, like Paul Wooten, who went on to study physics at Princeton. Paul had a calm, thoughtful, un-nerdlike manner, and could handle math problems that were opaque to me and do it without raising a sweat. Sitting beside him in class I felt like the klutz in a locker room of gifted athletes. It wasn't a total shock because I'd already had hints there were people who understood math in a way I never would because Dad was one of them. But he was an adult and Paul was my age. When he opened the textbook, he was looking into a brightly lit room, its walls covered with entrancing new symbols while I stood beside him staring into the blinding light, waiting for some dog to guide me.

Mr. Belge knew most of his students were smarter than him, too, but his view was broader than a simple hierarchy of mathematical aptitude. One day our class met with him immediately following an assembly in the school auditorium for seniors where the principal gave what was by now a familiar lecture on the benefits of a college education. This was unnecessary since most of the seniors came from families that expected them to go to college and a majority of these had the money to pay for it too. But the principal couldn't help himself. A former PhysEd teacher, a cliche on two legs, Mr. Cain had an obsessive preoccupation with college degrees, not for the intellectual experience they represented, not for the personal development of a student or the cultivation of the life of the mind, but to make money. Also to glorify his administration. Naturally, this made us all cynical. There were plenty of kids who saw college the same way as the principal did, but weren't so crass that they advertised it as their main goal. Or they did, and laughed, mocking themselves, while resenting the principal for seeming to turn them into his pets. Whatever the reason, the respect for him and his hypocritical piety for "education" was near zero.

After the assembly we went up to the second floor for our Advanced Algebra class and were sitting in our seats talking and making cracks about anything but Mr. Cain's boring, repetitive, venal speech, when Mr. Belge came storming in. He seemed to be preceded by his bulging belly, and swung his arm back to slam the door behind him. Something had pissed him off and his entrance grabbed our attention. He stood in front of us, belly-front, suit coat hanging open, and ran his hand over his blond crewcut as if he needed time to compose himself.

"What you just heard down there in the auditorium...it's a lot of bull--" he said and we all knew he dearly wanted to finish the word with the second syllable, but this was 1960 and speech was more polite than it is today. "Let me tell you something. If you want to make money, if that's your goal, the smartest way to do it is learn a trade. Become a plumber!" he almost shouted. He went on to run down the numbers on learning a trade and going into business versus going to college. "First of all, you won't be out ten thousand dollars in four years after paying for college," (again, this is 1960). "Second, after four years of plumbing, if you're any good at all you'll be *making* a good ten thousand a year." The fact that Mr. Belge looked more like a plumber himself than a teacher of mathematics gave his argument some bite, although all of us could see the inspiration for the lecture was a private one, not a rationally worked out lesson in economics for a room full of smart college-bound seniors. Mr. Belge might have had a brother-in-law who was a plumber and reminded him without mercy that his salary as a teacher was a pathetically small fraction of what the average plumber earned year after year, that his brother-in-law was in business for himself so he didn't have to suffer all the bureaucratic chickenshit regulations the Board of Education administrators inflicted on a public school teacher. Mr. Belge's argument had so much passion and energy behind it that he was still fulminating when the bell rang ending the abbreviated class. He had certainly gotten our attention; not because anyone left the room considering a career as a plumber. What we noticed instead was his defense of education as a means to something other than money. The idealistic earnestness motivating Mr. Belge was in its way a breach of decorum as great as Mr. Cain's venal promise of the income we'd reap if we went to college. It was a breach because Mr. Belge was expressing our hopes and ideals out loud, something we were careful never to do. It was hopelessly uncool, an invasion of our private territory. It silenced us to hear our secret hopes and ideals expressed so unequivocally, and to glimpse if only unconsciously that Mr. Belge was exposing to us his own beating and possibly wounded heart.

40. Bearing books

There was a blue canvas bag, a medium sized duffel bag, which I used to fill with heavy books like *Bible Stories* and Mom's *Winston Simplified Collegiate Dictionary*, five or six of them, and carry around the house. The purpose of putting the books in the bag and lugging them here and there was as obscure to me now as it was then, just as obscurely motivated as Dad taking all the socks out of his drawer, then carefully putting them all back in, and repeating the process several times, sometimes keeping at it until he was stopped. He did that when he was in his nineties and suffering from dementia. When I put the books in the bag I was eight or nine, with no brain damage from transient ischemic attacks. The books were totemic to me, for some reason. The knowledge they contained had some meaning, or would in the future, and the stupid exercise of lugging them from room to room in the house was a first step in our relationship. One day I would know what these books contained, or other books like them in heft and gravity. I would possess them, and they would in some way possess me. As they already did.

These books also played a role in another activity I didn't understand. I was constantly identifying some object as "the thing I liked best." This could be anything--a bag of books--although it was frequently something new, a toy, an item of clothing like a coat or pair of shoes, an addition to a collection of things which then expanded the "thing I liked best" to the whole collection, like marbles or baseball cards. Or more abstract categories like everything in my bedroom, rearranged into a new order poised to accomplish something. I didn't know what it might be intended to accomplish, and the novelty of this new entity that was liked best seemed to expire almost as soon as the rearrangement was done, as if it was all preparation, a process, not a goal.

The word for this emotional bond I formed with objects is cathexis. It surprised me to discover there was a word for it and I didn't learn the word until years later, so it was a long time before I knew this experience had earned a place in the dictionary. In my private world it wasn't just ownership, it was the way the bond with the particular object made the world complete, or constituted a complete world itself. By using the object, riding the bike, catching a baseball with the cherished glove, or wearing the boots or coat or playing a game with the wooden chess pieces on the antique marble board, I was admitted into this compleat world where nothing was wanting. It was like completing a puzzle in my brain which created a sense of integrity, a wholeness which included me. This abstract sense of completion was so palpable it was almost physical, or I should say the sense of being fully integrated was both physical and mental, abstract and concrete. I could savor this state for long minutes, sometimes almost doing nothing but enjoying it. As it took shape in my mind it brought with it a sense of timelessness. It was going to last forever. Then, as the spell began to weaken and fade, it shed it's beatitude and it began to seem as if the motley real world made up of undistinguished objects and lacking any abstract coherence or relationship to me was the way things really were, that this indifferent jumble of mere things had always been there waiting for the spell to dissolve and that I should have known the smooth, seamless, perfect state I'd been enjoying couldn't last. If it was centered on an object like my collection of Dinky Toy cars and trucks, they retreated into their unaffiliated selves and I felt foolish for ever thinking the mesmerized state of mind they created would insulate me from the fractured and splintered atomic reality we all lived in to the beat of a ticking clock.

41. Grammarian

The sanest thing in my life is this tone of voice. I come back to it now and then to remind myself of what I sound like at my most sane, when I'm most comfortable in my own head. The voice is also limiting. I can use it, cultivate it, encourage it, but I didn't make it up. I didn't create it. I grew into it by degrees. In some sense, it's natural. Or familial. I was the wee boy talked to by his mother. Her voice created mine. There were times I had to get away from the person who spoke in this voice too. The need to escape from the voice has led to some trouble. What I couldn't escape from was the feeling that the words had a structure, almost a geometric form, one I inhabited, as it inhabited my mind.

People have made smart comments about this quality of language. The most famous or first to come to mind is the one that says the French don't care what you do as long as you say it right. It's a backhanded tribute to their nuanced, high-wire grammar (along with a de rigeur swipe at their morals, rather out of date now, I would think). Another Frenchman, Montaigne, not known for facetiousness, said, "Most of the occasions for the troubles of this world are grammatical." That sounds like a clever way to say we fail to understand each other because we don't express ourselves well. My favorite crack about grammar is Mark Twain's: "A man who will use a parenthesis will do anything." Perhaps the inflated paeans to grammar show how deep down into the brain the wiring of language goes, and that's the real point. It seems to be very close to where a sense of self is recorded in neuronal connections. I'm only suggesting this as a possibility. I don't really know anything about neurology or linguistics. Knowledge of the brain has exploded yet we are just beginning to understand how much we don't know, a familiar pattern in our time, as Lewis Thomas said about our recent discovery of ignorance. When the sages say that it usually means they have recently learned a ton of things never known before but still lack ultimate answers. Or they have learned there aren't any. Something historic is happening in neurology. Some day we may be able to describe how grammar and the mental apparatus that organizes one's mind and grasps abstract structures are related. It's the prospect of learning things like this that makes me want to live, not forever, but longer, indefinitely. When my appetite for learning the answer to questions is sated, then I'll go.

Living through all the changes to come--the sea, the air, the earth and everything that lives here--is not something I want to do, however. As Robert Louis Stevenson said in his essay about the Marquesans, there is a limit to how much change a human being can absorb before the glue holding a self or society together dissolves. There will be calamities in the future which will be more than I can stomach, besides the tech innovations and social changes that can make me feel like I'm surrounded by aliens. My teeth aren't good enough to last that long either. My nieces and nephews and their progeny will have to accompany the race on the next leg. You can't escape your enclosing dates, you are pinned, stuck, slotted into a fixed and definite spot forever. We each have our own coordinates even as the unraveling scroll is unwinding away from our brief yellowing paragraph on the parchment.

42. Kiss

We were so young that the kiss was all of sensuality for us. The word says it in barely a syllable, from the catch in the back of the throat to the caressing hiss. No more than a sigh. But oh so much more. As Dr. Gillies loved to say, the most powerful sex organ is the brain. And the brain promises so much. Not the less for being limited, either, mutually submitting to the gentle pressure of our captive urgency. At that age, in those circumstances, at that time, much of our bodies are out of bounds. The fingers, hands, tongue and lips, are well within bounds we discover, and it's all so new. No other features are needed. We are teaching each other, learning touch and technique as we go, shyly guided by romance. The moments find us as we seek them out. Yet we aren't seeking anything, we're right here, consumed in the new moment. These are not acquisitive kisses. When our lips meet, the exchange, the reciprocity, enlarges upon itself when both feel it. Everything about this experience may be specific only to one, our sole mutuality. One can't kiss alone; another, a partner is necessary, and when both are engaged the moment blooms. There are so many reasons love like this might not happen but when it does it reorders all that preceded it; nothing else could have happened, nothing abstract could be so solid, irrefutable, such a fact. This is the wonder of the brain, that blissful organ, transforming the sensation passing through it into the death of irony. Revisiting the ecstasy, it slips away, transformed into silliness. Long ago and far away, in a different universe; how could we have spent so many formless hours kissing?

43. Disruption

"The sexual disrupts the form," wrote William Gass, "there is an almost immediate dishevelment, a proportion of events is lost." Like maybe the rest of your life is knocked out of whack and all proportion lost. Eros is it's own force, as urgent as gravity, and who needs to say it. Everyone knows--and cherishes or regrets--those secret moments when it wrapped you up in its arms and took you away, disrupting the form of your life.

"Eli, you're staring!" I hear someone scolding me, and the scene rises up like a stage set. The couple, kissing, might have been stage stars too, innocent as angel's breath in my twelve year old eyes. The setting could not have been more prosaic, a rehearsal room in the school where we were prepping for the All-City Concert Band performance. Most of the players were seniors and juniors but they needed someone to play the third trumpet and clarinet parts so there were a few of us virtual children mixed in with the musicians who were about to graduate into the world, a world where sex was just as big a deal as virgins imagined it was. What I was staring at--and until Lee woke me up from the spell I had no idea I was sunk in this state, so far had I been pulled out of my own consciousness by the exhibition of Eros in action a couple yardsticks away. The girl wasn't a girl, of course, but something else entirely--a local incarnation of Venus or Aphrodite or Elizabeth Taylor--and the guy was her erotic match, equal and opposite, angling his height around her, in dark suit and tie with a thick strand of his black pompadour dangling dangerously over his forehead. She seemed to be breathless, his kiss sucking the air out of her. That a man and a woman could do this with each other, become so entranced by performing exercises with their lips and tongues as if for a dental exam--yet their eyes were closed, they were swooning with need for this thing they were doing--and all around us were students getting their instruments ready, tooting a few notes, adjusting reeds, fixing their ties or hair or talking, making jokes. What the hungering Hollywood couple were doing was not allowed, that was certain, but they were doing it as plainly as I was staring at them, held in place helplessly watching their kissing. I was in some long, unnatural chute of time, my gaze as shameless as their lips. It was such a new thing to see, so far outside anything ever imagined. When Lee spoke to me, intruding into this space, I felt he was right to intrude, I had been caught doing something wrong. Yes, it was rude, boyish, immature, and all the more so because only my gaze was making their brazen breach of decorum visible. It was as if they were my precocious fantasy, so intense it had taken human form right in front of us all. If I looked away, they would vanish.

Lee grabbed my shoulder and yanked me around and a few moments later I saw the same guy, the one from the kissing couple, standing by an open instrument case, where lay a brazen trombone. He was joking with another player, now adjusting his tie, the strand of his pompadour still dangling but no longer dangerous, now only cute, and a few feet away ignoring him as if he had no power at all was the woman, holding a flute to her lips. According to Gass' formula, at this disheveled moment something else should have happened, some ripple in the flow of actions. But it didn't. Not there. The elements of the scene had gone off in their several directions, becoming the streams of events too minor to notice, the junk of daily days, ignored, forgotten.

44. Mom's boyfriend

"You could move over here or I could move in with you. Maybe I move in with you because your place is bigger."

There was a time, a crazy time, when Mom might have seriously considered Tony's proposal. Tony is my generation, a Vietnam vet, bitter and suspicious, also jokey, insidious, and alone. He bobs up in front of me, sarcastically smiling, his once buffed body beginning to soften, as if he's melting with age. Maybe Florida is doing this to him. Should have stayed in Chicago. Physically and emotionally wounded, he communicates with the world by complaining, arraigning everyone for their failure to help him. Just the kind of guy Dad probably feared his wife--my mother--would meet if she left him. Which she did for a while. That's all in the past, during her crazy years which can be explained rationally (if any emotion can be explained rationally). The crazy years were not her fault any more than Vietnam was Tony's fault. It was all history, ramifying in every direction, sticking to whomever it touched. His history was war, hers was peace. Tony speaks with that whining noise a human being makes when his life has gone wrong and it's too late to be fixed. He has no time. Happiness is not on the agenda, and it's someone's fault. What Tony's whine says is that his history is our history, so we owe him one, we're all to blame. Tony's whine multiplied by some number no one knows is part of the daily chorus America sings, in the Walt Whitman sense (except this is a song Whitman would never write), and we know Tony is there even if we don't hear his lone voice till he moves in next door. Now he is telling me what a shitty job my brother did of cleaning the old refrigerator after it was hauled into the carport beside my mother's "double-wide" in the mobile home park. These prefab little houses are actually *im*mobile; they never move again after being towed into place and set on blocks. As for the refrigerator, Walt did a nice job of an ugly chore when he cleaned it up, but Tony's best way of reaching someone is to piss them off. That's what he's trying to do to me now, with his squinty look and sarcastic smile. Life is a funny piece of shit, his face is saying. There is something boyish, even gentle, about Tony, although I can imagine him hurting someone (like Mom) just to see what it feels like, to see if it relieves his own pain. She told me she has seen him hit his cat with his cane to keep it from running out the door and wondered if he'd do that to her. Now he's telling me that the handyman Walt hired to do some work on Mom's immobile home wasn't handy. "He calls himself a handyman but he's not handy at anything." Tony is fond of this wordplay and repeats it a couple times, but I think he's getting the message that nothing he says is going to arouse my anger. Or make me laugh. I see through him. I don't dislike him, exactly. I feel sorry for him in a sardonic, Family of Man way. When it comes to Vietnam, I'm on his side. The country screwed itself and him in particular but I'm not going to help him. I feel the urge to hurt him, just as a warning not to touch Mom, although I can't imagine really doing it unless everything seriously escalated. The problem is having him next door to her. Why doesn't Tony with his Vietnam jones just move somewhere else? We all know Vietnam and what it stands for, the imperial American history handed down, one bloody hand to another, while the Country Club Chorus (all white, all male) sings "America," led by Graham Greene's Quiet American mouthing the words to cue the chorus who like the twittering orange-head hasn't bothered to learn the words. Tony doesn't know the words either, or won't sing them, and, as I say, he's entitled. He got shot in the knee, he told Mom, and whether it's true or not it is true his knee doesn't work right. You can see that. He is waiting for the VA to rebuild it again. Someone at the VA has it in for him, he says, and keeps pushing his name down the list on the schedule for surgery. It's easy to imagine someone doing that since he's so good at pissing people off. Tony's thin stream of bitterness pisses out of his organ of grievance every day, directed at anyone standing in front of him. He emits it just to keep his organism going. It's what he owes anyone who isn't helping him, the receipt for nothing.

Mom's friend Bobbi, who is ninety-two (Mom is ninety-five) teases her for having this suitor, this parody of a suitor. Does Tony want her money, or more of the food she cooked for him a couple times? It doesn't matter, she isn't available. "I had one husband, that was enough," she says with her quiet laugh, her eyes moving left, right, a little warily, wondering who has arranged to have Tony move into her life, trying to figure out how she is supposed to respond, since nothing, she believes, happens by accident. In her delusional world everything is arranged despite the disorder and she would like to know who is in charge, so as not to piss them off (a figure of speech she would never use). She likes having a boyfriend, someone who flirts with her, steals kisses. (He really does. When

Walt cleaned out her refrigerator and was about to toss out some old antibiotics she almost shouted at him, "Don't throw them out--I need them in case I have sex with someone.") What other forces intersect between her and him, her and Vietnam? "We already have an Asian man in the family," she says later, not really to anyone, long enough later that the reference, if intended, is lost. I hear it and understand but don't respond. The Asian man her daughter married is one of the things that unhinged her ordered life (how did it happen) and this oblique reference is the way her mind works, probably not uniquely but unlike anyone else I know. The way her mind works is something I pay attention to because there was a time when we had to have her committed, locked up. That sounds crude. It was done as gently as possible--but it was done, and it wasn't pretty. They had to grab her and muscle her into an ambulance.

She was mentally ill, there's no other word for it, driving to Canada by herself with housewares and silver and clothes and a television filling the backseat, telling people (like me) her husband was going to kill her. The docs declared her schizophrenic. An awful time, yet it happened, the way Vietnam happened, and like the war the wounds have mostly healed now except many people, both Asians and Americans, missed living their lives, and some like Tony are in their seventies, still looking at us with a squinty-eyed sarcastic smile, dribbling his helplessly earned bitterness on us like a dog marking a fire hydrant. If I try to describe who Mom was, and is, and what that difficult time was like when she was trying to make her own escape, I get into a tailspin of meanings and the impossibility of truly understanding anyone else. All that seems so tedious, so trite, so post-modern, to use a figure of speech I never use, and finally it's not a plunge into deeper comprehension but a trashing of the meaning of words and syntax and an abdication or renunciation of the burden of the gift of language--our gift to ourselves--to make things clearer instead of obscure and opaque and more confused than they are. The burden the gift puts on us is to explain life to ourselves, to portray us to ourselves, to make what we know understandable, sensible, graspable. Even when it isn't, even when final and definite meanings elude us. So what I can say about Mom? I can't really say anything in the same way I can say nothing about my own slough of depression way back when except that I survived, probably just by outlasting it, outliving it. That's all I know for certain. Mom survived her own mental illness too, in the sense that she is functional, although she is still crazy if by that you mean she understands life and the world she lives in under a spell of delusions that do not match commonly accepted versions of reality. Nevertheless, at her elderly age and despite her own fantastic and nutty yet sometimes wise and perceptive comprehension of things, she says she is happy at the assisted living residence facility she calls--what a long name for--home.

THREE

Hey, look--here's something I haven't seen for years, for more than half a century. Might have forgotten it entirely. But how could I? It's the greatest novel ever written, hovering there, just out of reach. It could be said that the creation of this rare, even unprecedented, work was made possible by another act of nature, by whatever fluke or genetic endowment created the author.

Colin Reese was aware at nearly every moment of every day that he was a wee man compared to the average. He was almost an elf, just over five feet tall, thin as a whippet, with thick dark hair worn long, horn-rimmed glasses, and although he shaved every day his beard was so heavy it looked as if nature was signaling that this wee man might have a boy's stature but there was to be no doubt about his gender or potency. Colin was brimming with testosterone traits. His manner was assertive, even aggressive, his speech emphatic, full of sound effects and loud strident laughter. If he denounced someone or something, as he frequently did, he'd sharpen his brogue--he'd been born in Scotland--trill his tongue up to a screech, and blister the victim with picturesque obscenities that felt like a slap in the face. Except it was also quite funny. All that exaggerated enunciation was obviously a performance. Instead of violence it led to a kind of respect, even wariness, from people he met. No one, at least in this country, had heard verbal fireworks like that before. To get something of his own back in return for having been made a man in miniature, Colin had fashioned himself--or had been fashioned by forces he couldn't really control--into an outsized character. Instead of hiding himself, speaking softly, avoiding noise or controversy or theatricality, Colin was all sound and fury, and would unstoppably quote you the famous speech the words came from, rising to a final quivering "signifying *nothing*!"

An essential part of Colin's performance were these quotations rendered in his trilling Scottish burr. He could hardly finish a sentence without a quotation set off by trills. If you were around him enough you realized that he didn't just have a modest stock of favorites, he was brimming with them. He may have made up a few too. Some were mock-aphorisms or folk sayings, but whatever the source they were all part of his repertoire. He often signaled one was coming by pointing his rigid index finger in the air, putting quotations marks into his speech with a pair of those trills, and zinging the semi-famous lines off his listeners. No one challenged the authenticity of Colin's quotations except humorously, or affectionately; it would have been poor sportsmanship. His vituperations were usually aimed at people who weren't present anyway and had made themselves egregious in some way that invited his wrath. Targets included politicians, official authorities, philistines, flag-wavers, and those obsessed with communists. Colin saw evidence of power trampling everything, everywhere, and was a libertarian as well as a socialist depending on the conversation. He'd read everything, it seemed; Dostoyevsky and Nietzsche were among his favorites, and a book he recommended to everyone was George Frazier's *The Golden Bough* which he also often quoted or paraphrased, along with Joyce, Eliot and Jung; also Robert Burns, of course, and the ubiquitous Shakespeare. He lent me his single volume version of Frazier's tome but even the condensation was a doorstopper and my wayward, undeveloped mind drooped over it. As if to balance these sages, he also recommended Ayn Rand whose ranting style appealed to Colin's suspicions of institutions, conventions, and entrenched authority. Her worship of power was embodied by individual supermen who challenged the state, so it passed Colin's test, even if her writing was hilariously atrocious, a parody of itself at times, a trait it shared with Colin's histrionics.

Performing his self-caricature seemed to enlarge him. His omnibus denunciations of institutions invested with power made Colin a citizen of the world instead of a cultural alien hiding out in an industrial city in central New York so unremarkable that it's most notable feature, the Erie Canal, had been dug almost a hundred and fifty years before and since been filled in, as if such a primitive, sluggish form of transportation--barges towed by mules--was an embarrassment in the modern era and should be buried underground.

Although I met Colin at the Elite Bar and Grille, and always saw him there, he drank very little. His intoxicant was talk. He drank ginger ale or seltzer and usually left the Elite early because he worked nights. He was a copy boy at the city's newspaper, assigned to the night shift. This must have been by choice because he had worked there for years and never mentioned any kind of friction on the job. A couple of us stopped by to see him one night and he had the run of the joint, swanning around the grimy desks, quoting and trilling his obiter dicta like a cicada as he tore off wire service copy, doing his job even while we distracted him. He took us down to see the printing press, the infernal machine smelling of newsprint and oil, the rollers humming and shaking the floor, the whole building trembling as it printed the latest news. The other workers seemed to treat Colin with an indulgent affection, just as we did at the Elite. Some of the regulars at the joint encouraged him to look for a job more suited to his abilities, but after seeing him at work it seemed as if he'd already settled into a career, or a way of life, that suited him. It allowed him his quirks, he lived comfortably at home with his parents (rarely mentioned) on his small salary, and had time to do what he wanted. Which was read. Also write. That was no surprise. Colin was the most literary person I'd met in my young life. The surprise came when one night he told me what he was writing. "It's the greatest novel ever written," he announced with a broad trill. "The greatest?" I said. This was my first experience with the grandiosity of a writer's ambition and I was too young to know it was often in inverse proportion to the number of words the person had put down on paper. Colin and I were sitting at one of the booths in the ironically named "Poinsettia Room" at the Elite. It was a slow night with no one else in the long, darkly wallpapered prune colored room and only a few customers were at the bar on the other side of the wall separating it from the line of booths. When it was just the two of us together, Colin modulated his act somewhat, although he never strayed far from it. Now I was smiling, not sure how seriously to take "the greatest novel ever written."

"What is it about?" I asked, expecting Colin to deflect the question, but he plunged in, not explaining but performing.

With nearly his first phrase I was lost. There were no handholds for my curiosity, nothing to visualize, no character or situation to follow. What Colin seemed to be describing were the concepts underpinning his great work and the effect they would have on those who would one day read them in their fully realized form. He detailed the many sections and interwoven themes he was going to need to be able to say it all.

"How long is it--or how long is it going to be?" I asked.

"Oh, a thousand pages. At least," Colin said.

"How much have you got done so far?"

But Colin had only paused to catch his breath; he ignored the question and was off again. His monologue shifted into subjects that may or may not have been part of his great screed and were swept up in his brimming enthusiasm. When he finally wound up to what sounded like a conclusion, I knew nothing more definite about his opus than I had before. Only that it was a work in progress, very long, and meant to encompass everything, all the themes of the flawed and mighty human adventure. Colin said some day he'd show it to me which seemed to hint he was far enough along that someone might be able to sit down and read it. But he didn't ask me to read it, not even a chapter or a portion, which I assumed was because I was too young and too unlettered to provide an intelligent response.

I only asked him about his great opus on one other occasion. This time his answer was terse and grudging, almost as if he didn't know what I was referring to. So, I figured, his novel was meant to be secret. It was not a subject for ordinary conversation. I imagined the fat stack of typed pages slowly rising toward completion and didn't doubt it existed in some form. I didn't begin to wonder about it until some time later, after I'd met a couple other self-described authors or artists who said or hinted they were creating a work that would stun the world, exceeding anything ever done before. At the time, it was important to me not to doubt Colin. I believed in the transformative power of art. My state of mind may have been a delusion, sustained by hopeful ignorance, and part of the delusion was siding with Colin, cheering him on against a world which more likely saw him as a ridiculous freak instead of a uniquely gifted, under-appreciated poet of the human spirit. I didn't realize until I heard my question dragging his opus out into the light again how much the ambitious work needed secrecy. Or invisibility. What he had described for me that night at the Elite was probably his actual opus--his great "novel" was this performance, the quivering finger jutting into the air, the cascade of ideas that sounded as if it could go on through the night and into the dawn and yet another day until it had portrayed every human emotion and every idea with an example of every human type, so when he finally wound up to the fire and brimstone conclusion, the trilled r's rolling and spitting off his tongue like bullets--there could be no doubt the end would be apocalyptic--a reader, a listener, would feel as if the scroll of life would never again be so magnificently and sublimely and thrillingly and consummately recorded, and these unreal, overblown traits were a certain proof that Colin's novel to end all novels could only exist as a dream.

2. Hot blue dot two

We don't really know what's out there, and when I think that cosmologists like Adam Frank have assured us there must be something, maybe a whole lot of somethings, including a few occupied exoplanets out there orbiting their own suns, it starts to seem that getting too hung up on what's back on terra firma means I'm stuck in the slow lane--you could almost say the no lane--while anyone with any imagination or ambition has already sent their mind into space even if their bodies can't go yet. Things will accelerate soon.

Sitting out here on my pyramid is sort of symbolic of this current halfway or partway state. What could represent the ancient earthbound world more graphically than this massive tetrahedron of stone? It may also represent the direction we're going without knowing it. Not forward into some brave new space/time but straight back to the beginning, I mean. That's if we're the frog in the pot of slowly heating water, being cooked without knowing it until it's too late for our overheated seas, soil, and air to recover. What we need to do is designate a particular day, today would be a good choice, as the day global warming began to be taken so seriously we dropped everything else to deal with it. It's apparent the human mind, I mean the collective one, the hive mind, is unable to make that decision. So we, the frog, will stay on the stove and cook. Ever since we got fire it's been a theme among skeptics to question the blessings of tools and technology by asking if our inventions will improve our lives by extending our power, taking us to unimagined places and discoveries. Or will they destroy us? The fear that we're too clever for our own good must be as old as Prometheus. Maybe we're going to have to go through some unspeakable Armageddon and hope for a rebirth before we can take a step into this new age (that's what Tom thinks). The process of disintegration and rebuilding would be a very long one. With suffering on a scale no one has seen yet on earth and the scary possibility that we might get stuck in some dead end replay of the Dark Ages instead of emerging into Enlightenment 2.0. Our first chance may have been our only one. So drop everything, run out in the street, and...do what?

Celebrate Earth Day?

That's where all this ends up--making lame jokes.

The first Earth Day was 1970. Half a century ago. We've had plenty of time to do something about trashing ourselves. Why didn't we? Because the boring, perpetual, ineradicable problem is...

I'm not going to say it.

When you write something like this you feel stupid. As in: how stupid can you be to think earnest words make the slightest difference? Giving sermons! As Robert Louis Stevenson said on that subject, "My father and I both liked to give sermons, although no one ever said either of us enjoyed listening to one."

3. The original artist

The calendar photo of the cedar waxwing snagged my eye as those nature porn pictures are meant to do, and for a moment I was stunned by its beauty, as if I'd never seen anything so perfect before. The subtle shadings of soft brown color, the black and white accents on face and wings, the neat dabs of red and yellow, the overall balance created a composition that was completely fulfilled and could not be improved. My response was emotional too. I loved this living thing and wanted somehow to make it mine. The crest on its head looked rather rakish, as if the visual display was a conscious achievement by a creature with a highly developed sense of style, just waiting to hear your praise. The crimson wingtips that give it its name are like drops of blood, a woman of the night's brazen lipstick and nail polish, another sign all this is intended and not wantonly wasted on some common bird. Yet the silky tan body is plump and comfortable as a burgher with his easy chair and TV, and the ornamental details are as tasteful as they are striking. Nothing is overdone. The total effect is so composed and arresting that the first time I saw it I assumed it was an artist's production, not a real bird. But the cedar waxwing, like its close bluish cousin the Bohemian waxwing, is not made up. If there is any esthetic intention in their coloring, we don't know what it is. Which leads to the question: If nature doesn't create these effects intentionally, if it isn't guided by an awareness of what it's doing, how does it do it? Or to ask even more simply, who or what is making these artistic choices. Whether it's conscious or not, how does the bird design itself? Are these effects really the products of evolution?

Richard Prum, a Yale ornithologist, has suggested part of the reason. Prum says that biologists have been interpreting nature using only half of Darwin's theory. Besides "descent with modification," the slow process of paring our traits down to those most adapted to survival, there was another leg Darwin's theory stood on, one which has been almost forgotten. This is the effect of mate choice on evolution. If females have some choice over which male they accept as a sexual mate, their choice will directly affect which traits are passed on to the next generation. (Males too, obviously affect which traits are favored, but this is not the novel part of his argument.) It may be, says Prum, that phenomena which have always been hard to explain using the theory of natural selection, like the flamboyant beauty and outrageous uselessness of a male peacock's tail (Darwin said it gave him a headache to think of it because of the way it challenged his theory) can be explained quite satisfactorily by regarding the tail as a product of females' collective decisions over time about whom to mate with. They favored extravagant tail feathers, the more extravagant the better. If Prum is right--and his argument is much more profound and subtle than my sketch of it--it can explain why a trait would thrive that is esthetically pleasing despite lacking any obvious benefit in making a species better adapted to the rigors of existence. It may be that by studying enough birds in the wild and crunching enough data biologists can show how such striking and subtle effects could grow out of umpteen breeding decisions, although that still stops short of explaining where nature got its esthetic sense in the beginning. It seems that some other force has led to the expression of esthetic values in design, color, and form we see everywhere in nature. Perhaps this is a force or influence related to structure, like symmetry, or some other underlying quality not visible on the surface but present in all living things.

Since no one has yet defined a formula for life, we can't deconstruct it to see if it includes some esthetic generator. More likely it would take some form we wouldn't recognize anyway. Not directly or immediately, since we do know how jealous nature is of her secrets and how well-disguised they often are. I would like to know the answer to the question about the source and essence of nature's esthetic sense as much as I'd like to know anything, all the more so since it seems likely that the answer will be woven into some very deep insights into nature's nature. At my age I'm unlikely to know this by the time my eyes close finally. So I have to content myself with pondering the mystery. Since there is always a new, deeper mystery nested in any answer we discover, I can't, in good faith, complain. The answer to the last question, if there were such a thing, would only be another question, and to make myself feel better I picture the mental pursuit in the shape of a circle or a Mobius strip and I have to be content knowing this motion--going around on a circle, or enjoying the Mobius roller coaster trip--is as far as my mind can go.

4. Offensive ideas

"I am at present castrating my work..." David Hume wrote to a friend in 1737, "...that is, endeavoring it shall give as little offense as possible." In saying this, Hume also admits with his usual candor: "This is a piece of cowardice."

Why is he doing it? Carl L. Becker, the historian who quoted this, says, "I think Hume's real reason for soft-pedaling skepticism was the feeling that such negative conclusions were useless." Becker also quoted Hume saying, "A man has but a bad grace who delivers a theory, however true, which...leads to a practice dangerous and pernicious. Why rake into those corners of nature which spread a nuisance all around?" So it may be cowardly but it may also be better to leave some things unsaid.

This is quite surprising to read, coming from one of history's most notorious skeptics and truth-tellers, who still sounds relevant today because of the cold eye he casts on self-deluding enthusiasms. I think Hume's misgiving about the risk and cost of "spreading a nuisance" has its roots in the breach between nature and the quite unnatural world we've created. Nowadays we're all raking into the corners of nature in ways that spread a nuisance into the minds of many people, often upsetting them so much they can't live in peace. The chief example--the epitome--of raking nature to uncover its secrets must be Charles Darwin presenting us with *On the Origins of the Species*, to tell us the unpleasant news about where we came from. This was a hundred plus years after Hume confessed he was busy "castrating" his own work so it wouldn't give offense. Darwin had delayed for years announcing his theory of evolution--castrating himself all that time--because he knew if people understood what he was saying it would give enormous, unprecedented offense. It would upset people, including his wife, so profoundly he'd be considered a threat to civilized society, which is exactly what happened, and it happened even though he was no rebel. Darwin's description of evolution, no matter how humbly presented in the cool language of science, could not help undermining the common human assumption that we were anointed by God to be his special creatures. If people understood what his theory was saying it would shake everyone's mind so hard no one could ever look in the mirror with the same confidence again. (Lady Astor famously said about *On the Origins*, "I hope it isn't true, and if it is, I wish he hadn't said it.") We've had to live with this "nuisance" ever since, and it feels now, looking back, as if Darwin's *Origins* blew up the rock, the foundation that all kinds of assured beliefs rested upon. Only a few years after Darwin undermined Victorian complacency, Nietzsche, the son of a Lutheran pastor who lacked the sociability of Hume and also lacked Hume's sensitivity to our feelings, announced, "God is dead." It was shocking to hear this stated so rudely, and again people could only hope it wasn't true. But momentum was all on Nietzsche's side, and as the twentieth century unreeled, the magical power which religion had jealously owned since Adam and Eve, was abducted, one discovery at a time, from the temple to the laboratory. It became apparent from these discoveries, shining so brightly in the light shed by the enlightenment, that what Nietzsche had said was not just a provocative wisecrack, the kind of thing an adolescent might say, but that a supernatural Deity was a myth we would have to live without. Another hundred or so years after Nietzsche's pronouncement, the poet Czeslaw Milosz wrote:

"If there is no God,
not everything is permitted to man.
He is still his brother's keeper
and he is not permitted to sadden
his brother
by saying that there is no God."

Milosz expresses Hume's tone exactly. Whether he's disguising his own faith or protecting what's left of the historical relic, or both, is in question.

After the world wars and horrors of the twentieth century, we could no longer say "not everything was permitted to man" except as a hope, or ironically, tragically. When viewed from a pyramid in space, Milosz's gentle scolding sounds quaint, as much as it shimmers with beauty and morality. That's the aspect one can see most vividly from the pyramid, and from here it seems possible the residents of earth might find a way to resurrect their God as a metaphor. Under such a transformation, the great concept might even grow in strength, relieved of the absurd burden of proving it was real instead of acknowledging the radiant creation was imagined. Why can't we agree that if the human imagination can come up with an idea like God it has the potential to be an inspiring force, greater than any one of us. (There is a downside too, unfortunately, which will occur to anyone, and that is the monsters of evil our imaginations can also conceive. A reminder that no yin comes without its companion yang clinging to it.)

Since Hume's decision to spare us the worst he'd discovered about life, so much more has been learned, both cheering and depressing, that the thought of editing or castrating your work because it might upset people sounds quaint too. It is certainly not something a scientist would do. More likely the scientist would rush to publish alarming news because it would bolster his or her professional reputation. A philosopher wouldn't censor an earthshaking discovery either, although it's hard to imagine anything a philosopher could discover that would have such an effect. Science owns all the surprises.

Ever since Darwin, we've had to learn to live with discoveries and innovations that force us to re-examine everything. That brings us up to the twentieth century and the power to blow ourselves up, a threat that looks like a permanent addition to the human condition, as does the twenty-first century's requirement that we manage, instead of just using and consuming, the earth's resources if our civilization is to endure.

5. The big room

Gus was gruff, not talkative, an old man now who'd heard it all and didn't listen to most of the bullshit flung in the air, along with cigarette smoke, on the other side of the bar. But one noisy night when the Elite Bar and Grille was jammed with drinkers, talkers, arguers, and laughers, and he was busier than he liked to be, much busier, he walked down to the far end of the working side of the bar with a glass and mixing canister in his hands--he hated mixed drinks (have a shot or a beer or go home) and out of the din he heard one of those earnest youthful voices barking out some profound observation. It interrupted Gus's attention so completely that he couldn't remember what he had come down to the end of the bar for. In his confusion, he stopped, looked blankly at the mass of young faces, and barked back at the philosopher-kid:

"Life is a big room--ex*plore* it!"

It popped out of him so unexpectedly it surprised him too, possibly surprising him most of all, because unlike the know-it-all students and theater people and regulars and assorted misfits and boozers showering each other with such important pronouncements, Gus didn't arrogate to himself the privilege of commenting about all of life and all of its meaning or lack of it, and sometimes, hearing so much blarney on his side of the bar, broadcast so loudly and confidently with drink, he wanted to tell them all to shut up. In Gus's world it was the priest who made those pronouncements and when he did he usually spoke from the pulpit and his voice was tinged with the sanctity of ritual and tradition. He wasn't making a wisecrack. Gus was an intelligent man but he never assumed he had the right to make pronouncements about life itself, the life we shared with all Creation, a strange and wondrous sacred thing so much greater than any of us that to speak of it in your own voice was risking blasphemy. To hear himself sounding like one of his young customers was a shock. But to hear them all suddenly stop talking, and look at him, and then spontaneously cheer in unison was funny, so Gus laughed with them--he couldn't help it--and at himself too, perhaps even glimpsing for the first time how much affection his customers felt for the grumpy old man behind the bar.

6. "Read all of Faulkner!"

A collage has edges even if it doesn't have a formal beginning, middle and end, and one of the edges is my friend Roland. He is sitting eternally on a stool at the Elite, which itself has been swallowed up in history under the euphemism urban renewal. His pleasant baritone voice can often be heard in the humorous, ironic din produced by the crowd of patrons on a busy Wednesday night when a number of grad students from the university often joined the theater people from next door along with a sprinkling of regulars, mostly working class guys who hang out together at the Y to play handball and grew up with Shin, bartender and co-owner of the Elite with the older Gus. On busy nights this blend of humanity, mostly male, fills the joint creating a jokey, wisecracking spirit. What can also be heard two or three nights a week is Roland's laugh, a big soaring eruption that comes in waves, inviting you to ride along with it.

Roland was friendly, almost ingratiating at times, a trait that came out when we first met because he had an unusually strong sense of place and didn't want to offend anyone else's sense of place by intruding into it at the Elite. He felt this particularly in that locale since sitting there alone at the bar he was an outsider, a newcomer. With its motley clientele, it was not a neighborhood joint, places which can be frosty to outsiders, but he felt his own strangeness. He wasn't from central New York, he was a Californian, a native one he hastened to tell you, third or fourth generation, and when he said it you could hear his pride, his possessiveness of San Francisco, the Bay area, and the Sierra mountains, their beauty and history, and his place in it. Had the geography been reversed, would he have felt so proud and possessive toward central New York and the Finger Lakes, the Adirondack Mountains and Lake Ontario as he did toward California, despite how much they differed in scenic effects and as a setting for one's own biography? It's quite possible he would have since Roland's attachments to whatever he liked or disliked, approved or disapproved, owned or disowned, were always strong, even fierce. He'd have loved my unlovely provincial, industrial hometown, transforming it with his devotion into a place anyone would be proud to have grown up in. (As Shin and his buddies from the Y felt about God's country, as they called it, before adding with a laugh, "if you like snow.")

Roland taught philosophy and literature at San Francisco State College and was spending a year at Syracuse University on a fellowship. He had published a book on education and the humanities which was used as a text at more than a hundred colleges and universities throughout the country, he told me both proudly and diffidently, as if it wasn't his real ambition. Besides his promising academic career, he was a serious classical pianist and had also played and sang in clubs in San Francisco while working his way through college. He was a knowledgeable and reverent guide to the Sierra Mountains of northern California. His heartiest laugh, a Henry the Eighth guffaw, expressed his interest in what used to be called female companionship. The phrase sounds all wrong today, which may be at least one too obvious clue to Roland's increasingly poor fit in a world whose changes, except for civil rights and the ecology movement, he was going to resist, in his fashion.

On that first occasion when our trajectories crossed, Roland appeared sitting by himself at the bar of the ironically or at least ambiguously named unpretentious beer joint down the hill from the university. He is smoking a pipe as if he doesn't enjoy it (trying to quit cigarettes--in a bar!) and noticing other people around him discreetly, to be seen but not to intrude. He called himself "Rollie" then and if he'd been on his own turf instead of ours, he might not have been so accommodatingly friendly. The reversion to his full name came later, as if he'd realized he was letting the world be too familiar with him, a penalty for being so naturally affable. Perhaps his ready fellow-feeling countered a deeper down wariness and even a trace of malevolence. The reason for that--a possible, plausible reason for it--comes later. Right now Roland, Rollie, is introducing himself. I'm working behind the bar though that's Shin's usual province and mine is to wait on customers in the booths in the adjacent Poinsettia Room. Shin has gone off to do an errand and will be back in a few minutes. Roland, with that anomalous pipe (which will never be seen again), has introduced himself. He's a big man, almost bear-like, and leans forward a couple degrees, walking with a peculiar shuffle, so he doesn't look athletic. Later I learned he was a better-than-good athlete, a pitcher and light-heavyweight boxer in college, and fiercely competitive. (There's that word again.) He was also a man with physical courage. Although--and this is strange--his handshake is soft. It feels like he's wearing a mitten. I can't explain this disparity except that he disdained macho posing and giving someone a death grip handshake would have been way out of character. Yet if you challenged him, he would fight. Almost gladly, as if it was sport. In a word--if there had to be only one word--he was a warrior, a genteel warrior for whom life was, despite his affability, a mortally serious contest of wills and ideas.

We only saw each other at the Elite, often on Wednesday nights when I didn't work and could join the self-parodying Wednesday Night Literary Club for irreverent and occasionally serious blather. Roland and I were almost instantly simpatico in the way that friendship can be a species of romance. He was ten years older than me, fluent in philosophy and literature while I was a mostly ignorant dilettante so the relationship was unbalanced, but there were some core values we both revered. Roland hosted a craving for excellence and describing this gave it the ancient Greek name *arete*. Literally it referred to a mountain peak or ridge, a sierra, which chimed nicely with the range of mountains he had spent so much time hiking and camping in. As a metaphor, the goal they represented was something ineffable, something one could strive for but was ultimately unachievable. This kind of lofty conversation in the dingy Elite usually produced at some point a mountainous guffaw from Roland, which spread with ironic and lusty contagion. The contrast was funny, yet we were serious too. Even if, in my case, I had no real idea what it meant.

It was a great year. I met Maddy, fell in love (we were all a little in love with her), and Roland acted as a senior sponsor of our romance. At the end of the year on a basking spring night Maddy and I and another friend of Roland's drank sayonara together at the Elite, then drove to the airport to put Roland on a plane to fly home to his wife and children. He gave me a copy of *Light in August*, inscribed "with fearless high hopes for his future." He didn't imagine me half a century later moving a compost heap, nor would I have ever thought that one day half of Roland's heart would stop working and he'd have to beg for alms.

7. *Dead man walking*

Roland may have had too many talents--scholar, writer, musician, athlete, and later polemicist and radio and television commentator--and perhaps the multiple talents meant there was always an alternative tempting him away from whatever he was doing. His ability as a pianist wasn't insignificant but it was minor in the diabolical way that music can tease its gifted acolytes into believing they must have been singled out by fate for celebration. Instead they learn they are blessed with an unusual talent which is shared with a surprising number of other unusually gifted people.

The sensibility Roland developed in learning the standard classical repertoire may also have led him as a writer into the temptation to create something monumental. First a novel, then a movie script, then the television series he called "Spiritus Mundi." Like many before him, he wanted to create something timeless and great. Like a pyramid. Well, no, not a pyramid. Yes, exactly that. Something timeless and great. After the many years of trying to do this in several forms, to impose his lofty vision of human existence on the world, two chambers of the blood-pumping organ quit, they unpoetically quit. On the phone he described walking into the emergency room and telling the nurse he thought he had the flu because he'd felt fatigued for days. He hoped they could give him some medication and he'd be on his way. She put a stethoscope to his chest and said, "You're not going anywhere, my friend." My friend! How could half his heart quit working yet he was still walking around, driving, doing things? A few months before he'd been hiking in the Sierra! Well, that's why you feel so fatigued, the nurse told him. "You're lucky this didn't happen in the mountains." His heart was working awfully hard, and not very efficiently. Then, while he was right there in the hospital, it stopped. They put the paddles on him and shocked him back to life, like Frankenstein's monster. It was an agony, he said, waiting each time as the nurse said, *Ready? Here it comes...BOOOM!* And another jolt revived the stricken organ.

Roland calling to ask for money was also a shock. He asked for a loan of five thousand (it could also be a gift, he suggested, and if I could spare more it would be welcome). I told him that besides not having that kind of ready cash, I had cancer myself. It was not a good situation. I had my own health problems and expenses with looming and unknown future needs. As a free-lancer I had no income when I was sick. So we were both needy. But not equal, it seemed, in need or worth. When the doctor came into the exam room, Roland said, he recognized his patient as a multi-talented local star and said, "Here's the genius." To my sensitized ear, he seemed to be telling me this as if it gave him a claim on my resources. He was the genius, I was the bank. Since I didn't have any money I could send anyway, maybe I was just feeling defensive about turning him down, chagrined I couldn't help. It was all very awkward; a tangle of broken faith in several ways, drawn out over decades.

The day after our phone call I felt heavy with regret for not being able to rescue Roland, or at least help him some way. Something like this had happened before, years ago when he was in much less desperate straits. He hadn't asked for nearly as much money and as it happened at the time I was flush so I sent him a check. He sent back an elaborate written agreement with a payment schedule. I understood that this was his pride, making sure I knew we had a quasi-legal agreement and he hadn't been simply asking for a handout. Roland was a man of good faith who kept his promises and paid his debts. Of course he was, you knew it instinctively when you met him. He was a middle class guy who kept his word. After the first payment came due, which he dutifully paid, he fell behind and it soon became tacitly acknowledged between us that he'd pay me back when he could. It turned out to be such a long time that finally, out of embarrassment for him, I wrote and told him since I'd written off the loan long ago why didn't he do the same and we'd just forget the whole thing. Since in some organic sense, neither of us was precisely the same person those several years later that we'd been when I gave him the cash, he was able to exonerate the person he no longer was anyway for defaulting on a "loan." The problem now was that in what he said asking for the new loan--perhaps to help him justify asking for money--he wanted me to know that at least in his own social world he was considered a figure worthy of tribute. Well, fine, I thought he was a most unusual person too which is why our friendship, particularly in those first few years, had been such a rich experience. But life is a crapshoot with scant sense of justice, even for the rarest among us, and if it deals you a sudden mortal illness all bets are off in a heartbeat. So to speak. I just didn't want my invisible tombstone to say *Here lies a bourgeois television producer who chose the safer road to middling achievement over his soul buddy Roland's high-risk, all-in ascent toward*--and here I had to stop and go look up the Greek word for excellence he'd introduced me to on one of those enchanted nights long ago in the Elite Bar and Grille. My memory of the meaning of the elusive word was accurate even if the word itself had slipped away. "*Arete* means excellence of any kind and was ultimately bound up with the notion of fulfillment of purpose or function: the act of living up to one's full potential." Reading this now was as humbling as it had been inspiring to hear Roland describe it so many years ago. If we were both deluded--and, yes, thwarted--it was in our separate fashions. Among the things we had in common which was only now fully apparent was a gift for making things hard for oneself. Now, before Roland could complete himself according to his personal ideal, his body had failed him, just as mine was doing to me less dramatically (and I would recover).

There had always been an inequality in our friendship, right from the first night in the Elite.

Now that our bodies were failing us and our personal clocks were running down, it was not surprising that certain not entirely comforting truths underlying our long friendship were emerging. The question was why did it take so long to come out? It was only because we spent most of our lives on opposite coasts and saw each other so rarely that the crevice between us--which I called "Vietnam"--hadn't swallowed our friendship long before this. Even so it does seem a wonder since the war was so tragic and divisive and opinions about it so defining. When Roland and I met, a few American "advisors" were over there and their presence was not a hot topic. Mostly we avoided subjects like the domino theory and American imperial adventures. We communicated on a shared higher plane. It sounds strange to say that. What's higher than life and death, than the subject of war (though neither of us was ever in one)? This sounds like a willful moral blindness on my part, if I believed what I said I believed. But my friendship with Roland was like no other. It was as if he and I were members of some institution, like a Senate, where we could disagree about particular issues, even war, without tearing down the forum in which we met. Because the forum was the sanctuary of a higher ideal we both believed in. This gets closer to the feeling between us, yet in the end there is going to be no way for me to explain this, not to my own satisfaction. It must stand as a contradiction, unreconciled, unresolvable.

That our friendship had now suddenly come apart, or could no longer continue, shows the gap in understanding was human as well as ideological. Whatever the reason, the crevice had gaped in front of me when Roland asked for money. After we talked I thought, even if I had the money I did not want to subsidize Roland's current political ideas. The thought felt like a betrayal. How crazy all this was! For some reason that's what it felt as if he was asking me to do. Over the years, Roland's earlier enlightenment values had darkened so much that I no longer felt we were members of our own two-person Senate. That protected space had not turned out to be so capacious as I'd once thought. Some years before when he'd written me about a trip to the Middle East he had taken, he was contemptuous of the Arabs and their culture and lauded Israel and the Israelis. He was making a pro-western Judeo-Christian civilization argument, stirred by the anger aroused by Islamic radicals and terrorism. This kind of caricature of Arabs and Muslim civilization did not sound worthy of Roland. Years before, I thought, he'd more likely have told me how two great Muslim philosophers had helped save the works of Aristotle from oblivion hundreds of years ago, acts which were credited by some with igniting what became the European enlightenment. Perhaps that was too neat a sketch of the history of ideas, but I thought respect for knowledge was what we should be mindful of, not the grading of civilizations, which for me led directly into the problem of colonialism and thence to the crimes of our American past. The change in Roland was the result of our culture having turned into a place he no longer felt he fit. Standards had been degraded to accommodate diversity, America's posture among nations had become apologetic instead of assertive. We were even told to agree that there were no differences between genders. That gender itself could be arbitrary. Pursuing the ideal of equality had led us into absurdity. The politically correct rubbed against Roland's nature in several spots, no matter how much he believed in tolerance. Because he was outspoken and courageous defending what he believed in, his opinions may also have cost him his career in academia, although he would agree that his own projects always had priority over teaching or writing books on education. The affability I'd noticed the first time we met was still there but had been abraded and roughened in the long time since. Perhaps the clinching clue that we had tried mightily to protect our friendship from our divergent opinions was that we had never exchanged a word about Junior Bush and his shock and awe war on Iraq. Roland might have thought of it as a regrettable but necessary retribution for 9/11; it made me think of essays like Mark Twain's bitter screed on colonialism: "To the Person Sitting in Darkness," or "King Leopold's Soliloquy," or "The Slaughter of the Moros."

Besides the shocking cardiac news and the sense that something uniquely valuable was slipping away, there was something else unexpected and poignant about our last phone call. That was the story Roland told me from his childhood. Throughout the years I'd known him, this had been a closed subject except for select details, stingily titrated out. The dross of life was not for Roland to tell; it was something to push aside, to ignore, so one could move on to matters of significance. In Roland's fantasies he'd sprung from some goddess's ear. I'm talking about the kind of most private fantasies that live silently inside our heads, hidden even from ourselves; his sense of humor was too broad and irreverent ever to say such a dumb thing out loud, or even imagine it. His flesh and blood mother was a very warm and appealing person whom he teased with affection. What he described was a very different kind of incident: "When I was seven and my mother was having a rough time on her own, she put me in a home for foundling children for a year."

"What?" I said into the phone. "Really? She left you there for a year?" Roland was the oldest of three, I recalled, and she must have thought he was the one who could best survive in an orphanage after her children's father had left her.

"Yes, it was almost a year," he said but there were not going to be any more details. I knew that when he said it. Now I was trying to figure out what had spurred him to divulge such a telling incident from his young, most vulnerable years. It must have been the awful whiff of mortality from the cardiac crisis. Roland was not a vulnerable man. You could hurt him, of course, he had feelings, deep feelings, but he was fiercely proud (there's that word fierce again) of his independence, proud that he was no man's vassal and lived by his own rigorous code. His spinal willfulness was in a different realm than heart failure where he was very much something's vassal, as we all were.

"Have you told your son about that?" I asked. My question caught him sideways. Perhaps any question would have thrown him off balance since I suspected the revelation had come as a surprise to him too, a dubious gift of his confrontation with mortality and not something he'd intended to tell me, not when he had called asking for money, another humiliation. Or was it a plea for mercy?

"No," Roland said to my question, without adding (though I heard it), "Why should I?"

"Because then he would know something very important about you," I didn't say, since he hadn't asked why he should tell his son about what was probably one of the most upsetting, and formative, incidents in his life. It may also have been a key to his personality. Maybe not as determining as a genetic affect, but it might have been analogous to an epigenetic one, an event which switched on a trait that was latent in his organism, namely, the urge to fight. Life had hurt him badly, betrayed his childhood trust, and he wanted revenge. The few stories he'd told me about physical confrontations flipped through my mind. In the orphanage, big for his age, athletic, and armed with his grudge, he may have learned he was good at fighting, not only unafraid but enjoying a peculiar primal pleasure from the satisfaction it gave him to beat someone. Or, he may have been perversely scarred by a beating of some kind inflicted on him. Or both.

8. Arden's Journal

It's easy to imagine how Roland's mother might have spoiled him, treating him like a prince when she retrieved him after his year in purgatory at the foundling home. She felt bad about what she had done, of course, and how could she not? But life can be cruel even if you are white and at least originally from the comfortable class. Mother and son were back together now and she needed his forgiveness which he soon gave her. As full grown adults, they looked good together. They had the same bigger than average ampleness, the warmth and awareness, the wit and laughter, a look which in his case could suddenly flash with ferocity. She was only nineteen when he was born so they were close in age. His father, trained as a dentist, was long gone from the scene but Roland confessed to me once that the night his father died of his own heart disease, Roland was a guest on a radio show, sitting in a broadcast studio instead of at the hospital where, he chided himself, he should have been. It was when Roland himself was wheeled into the OR after his cardiac "event," that the doctor recognized him and said the words he said to me on the phone, asking for a loan, "Here comes the genius." Because of his appearances as a radio and television commentator on philosophy, education and civic values, as well as his playing and singing at local clubs and for musical events, Roland had indeed become a local figure, and with his myriad abilities on display, a genius. A few friends knew that he had also written a novel (which might never survive his severe editing), an ambitious movie script, several musical compositions, and a detailed proposal for a television series on the enduring values of western philosophy and civilization. This last was, in my view as a television producer myself, quixotic, and it had absorbed too much of Roland's resources (including money) for what was likely to be no reward. Which was how things turned out. His life was unfulfilled, by his own standard. Had he aimed too high or at the wrong goal? Aiming too high and at the wrong goal--if by wrong you mean too challenging, too unrealistic--was his nature. The only obstacle worthy of his ambition was one that looked insurmountable. That was the essence of Roland. When you're young, this posture is romantic. When you're old, it's...it is many things, but none is enviable or desirable if the end of life is fulfillment and happiness. Some people are fortunate in prudently nurturing achievable goals, while other lives, many fewer, like Roland's, seem to need an explanation.

Some years ago he sent me the first third of his novel in progress. I read it with great interest although it was hard going. It was hard because his sentences compressed--compacted--so much meaning, nuance, and resonance into each line, using layered, complex grammatical constructions along with arresting and insistent word choices. Plumbed to its depth, the style said life is a very particular challenge against forces often inimical to human sensibility and achievement, yet the only way to comprehend this depth is to risk all in the attempt. There is something operatic about this posture, something potentially tragic, and it is certainly not typically American. In Roland's hand it produced a heightened prose style you would never recommend anyone use, yet the elevation was an integral part of his purpose, and his art. The excesses were there because he was expressing what it meant to be an excessively aware and, yes, superior human being in the clangingly silent universe. The novel's main character was a woman (surprise), an astronomer (not a surprise--she studies the cosmos), and lives by herself on a houseboat in San Francisco Bay (as Roland once did). Her soulmate and intimate partner is an outdoorsman, an athlete and climber of mountains, with the assertive independence of a libertarian, and they share responses to life and music with a depth and sublimity the other uniquely comprehends. The woman, named Arden (as in ardent), has cancer and the treatment probably won't save her. There are long sequences, some of great beauty, describing the area of the bay around the houseboat, and as I said the writing throughout makes unusual demands on the reader. It is a kind of high wire act. There is a reward there, if you are willing to submit yourself to the text, as Roland himself might say.

What of the rest, the other considerations: were there characters, was Roland's novel alive? At times it nearly choked on its own eloquence (I told him my opinion tactfully), and the characters were versions of the author who existed in an imaginative realm of their own that was a kind of super-reality. I can't judge it as a complete work because I only read the first third. I don't think he finished the rest of it. I wish he had because I wanted to see if Roland's characters survived in the rare air of his art and what further experience and resolution he imagined for them. His form of written English was one I have never read anywhere else and in that sense he was original, as well as having something he wanted to say through the lives he gave his characters. So it demanded to be taken seriously. For the patient, motivated reader there was something there, something challenging, with a reward. At times the straining toward the higher elevations of human sensibility went over the top. As I said, there is not much breathing space on the page. Had Roland not been so wary, so private, so secretive even, he might have been able to inject some of his less heightened experience of life in a way that preserved what was unique in his manuscript, yet was in some instances throttling itself by excess. I tried to tell him this, again, tactfully, and suggested a structural change to move some of the action up earlier in the opening chapters. To my surprise instead of being offended he wrote back to say he thought that was a good suggestion. The style I don't think he could alter. It was him, unbending and difficult, but also inimitable. A self-portrait in prose, as distinctive as a font or fingerprint.

When he sent me the manuscript, he asked me to return it after I'd read it and enclosed a big, self-addressed envelope with postage already on it. I did. I didn't make a copy or keep a single page of it. I wish I had.

I do still have a piece Roland wrote which has a taste of his authorial voice. These are the "liner" notes to a tape he made of himself playing several pieces on solo piano. Describing Beethoven's Sonata No. 14 in C Sharp Minor, he says it has not outlived the tag of "moonlight" applied to it, then goes on to redefine his subject, something he often did: "Variously misinterpreted by pianists and listeners alike as some Keatsian swoon or silky melisma out of Debussy, the Adagio beckons in score as a drama of mind at the end of belief. Any moonlight here must shine on lunar desolation, and its tune is the melody of nihilism." Then Roland's clincher, citing the composer's own comment: "But in the same letter, Beethoven remarks on the strange value of one's life at odds with futility...though the cosmos appears to deny us the comfort of a decoded Answer, import radiates from our insistence that meaning somehow exists, if only in our rare and hungering wills or in the irony of nihilism. Our demand for meaning, not our fulfillment within it, is at the heart of our divinity. To ponder reality is to confirm our place within it."

Roland certainly made that demand of his own life but his attempt to fulfill this as a personal prophecy may have been anticipated by his warning that the demand was required, not the fulfillment. He might also have suspected that it was a goal beyond anyone, that its pursuit was his own abiding challenge and the condition of living his life. That pursuit, however it ended, revealed his courage, his fearless insistence on defining man at his most excellent. This sounds like a quest from some idealized past, nothing contemporary. The goal can also be seen as elitist and not properly an American trait, that it conflicts with the values of democracy. That may be true although it could also be--as it was in Roland's worldview--the juncture where excellence can be transformed into the spirit.

9. Triviality

Walking in San Francisco's Golden Gate Park with Kate's brother and his dog, we ran into Marshall. It was a small world moment. Marshall had gone to college with Roland, they were compadres, the oldest of friends, Roland had named his son after him. Marshall had had his own unconventional career as an actor, writer, and comedian in New York and Hollywood, and now he was back where he'd grown up. Kate's brother John had met Marshall because they lived in the same neighborhood and both walked their dogs in the park. Now Marshall had some chronic illness that was slowly killing him. His life was his dog, an elderly mixed breed shepherd exchanging sniffs with John's restless border collie. I'd met Marshall a couple times in New York years ago, when he was working as an actor and cab driver. When I asked him if he'd seen his old friend Roland lately, he looked astonished, as if I'd asked an odd question. "He threatened to sue me!" Marshall said, abruptly becoming histrionic. "He accused me of stealing his ideas in that proposal for a television series he'd written and he was going to take me to court if I didn't send it back to him by registered mail. He'd sent it to *me*--I didn't *ask* for it! He accused me of all kinds of stuff. He can be your friend. He's not mine," Marshall said shaking his head as if there was a whole world of things he no longer recognized and his old friend Roland was one of the most baffling ones.

Marshall was short, bald and rounded, with bugged-out Zero Mostel eyes; a very funny man and an almost too obvious Sancho Panza to Roland's Don Quixote. Yet this wasn't funny at all. I tried to place Roland's threat to sue him in the chronology--how long was this before the cardiac crisis? The vicissitudes of Roland's life seemed to be pushing any calculation away from me, and I couldn't figure it out. It no longer mattered. Roland was a multiply talented, complex man who made excessive demands on himself, on life, and sometimes on those around him; now, near the end, perhaps overcome with frustration and his own bafflement at how things had turned out, he was becoming somewhat deranged. He suddenly seemed like one of the original dead white men whose careers lay on the pyre of change.

Yet if that's who he is, or was, it made him merely a sociologist's dummy, and that's wrong. Roland was an artist, a musician and a moral philosopher, trying to express what it is we live by and to what end. His spiritual guide and soulmate was Marcus Aurelius. If that sounds inflated, well, he had the courage not to deny it. Trying to express this in his own work he confronted the problem at the heart of our middle-class existence: triviality. Once we become comfortable, the human animal's triviality is so exposed it can make us ridiculous. Thinking about Roland's fate--and that is the only word for it--I sense a whiff of the tragic and hear a quotation from antiquity: *Whom the gods would destroy they first make mad.* That is not a trivial outcome, nor is Roland's compressed, over the top, 44 ounce Louisville Slugger prose style trivial. Despite his humor and affability--and, yea, his irreverence--it accurately reflected the spirit of the man holding the pen. He was not someone you had met before. His uniqueness was expressed by his other-world style. The other world was the classical one. He'd have been quite at home with the Athenians. Not the real ones of two thousand plus years ago but the ones bred by history and imagination, still so alive in their own words.

10. The compost temptation

A compost heap has a strange attraction for certain people. It may be a guy thing. Philip knew a woman, a theater colleague, whose boyfriend was a fireman at the firehouse that serves Broadway's theater district, and he started collecting kitchen scraps during his shifts on duty to take back to the compost heap at his home on Long Island. But he had a rival, there was competition for the fire station's garbage; another fireman in the same station also had a compost heap he was cultivating and he wanted the kitchen scraps too. They were civil about it and divvied up the trimmings from the vegetables the firemen ate while at work, but it showed how avid devotees can be to get their hands on any available green waste to batten their heap. Certain guys--and perhaps women and children too--seem to be drawn to the subtle pleasures of piling up leaves, hay, vegetable trimmings, grass clippings, and lime and ashes, just to name the raw materials that go into my heap. It also gets the soil clinging to stalks of plants harvested from the garden, soil which adds the bacteria needed to break down all the cellulose and other once-living green matter to turn it into fertilizer for the garden.

Maintaining a compost heap is one of the most mundane, least thrilling activities I can think of. Yet I would not give it up, or trade it, for anything. In that sense it's like gardening itself or, now that I think of it, walking the dogs. Dull, unexciting, yet strangely satisfying. Besides having to compete with your colleagues at work for vegetable scraps, the heap makes other demands too. It needs to be turned. In my case, because we live in a rural area and have a large yard, there is plenty of space for three heaps. So I work on an annual rotation and only turn the heaps once a year. The dark, granular compost in the third, the finished heap, measures roughly eight by ten feet and is about three feet high. When we're ready to plant in the late spring, I shovel a load of it into a two-wheeled cart and roll it to the garden. Tomatoes, brassicas, squash, peppers, asparagus, strawberries, most everything gets a generous shovelful to nourish the seedlings and growing plants. It also slowly improves the quality of the soil. Over the course of the summer, I spread compost wherever it can help, since I have to move all the finished "black gold" to clear a spot to continue the cycle of heaps. Sometime in August or September, I begin shoveling the next pile of compost--the one in the middle--over the fence to the spot recently vacated by the stuff we've been using this year. By now I'm working against the clock, or the calendar, because once I've moved the middle heap, I have to begin shoveling last year's pile of once-living greenery into the middle location, to get ready for when the leaves fall. They will be raked up and used to start the first heap all over again.

Is this efficient? Does it make sense? Of course not. It is time consuming and as I get older I wonder how long I'll be able to shovel all this compost. Every year I tease myself with the idea of hiring someone with a small tractor and bucket loader to come and move it for me. But I'm strangely jealous of doing the job myself. The process of moving the heaps is part of my relationship with the compost. It can be made much faster than I do it. My process evolved simply because I had plenty of space and a lot of material to convert. People in cities need a speedier process with more frequent turning and a finer tuning of ingredients so the heap begins "working" quickly to break down the vegetable matter. The times my heap begins really cooking can be amazingly gratifying. In the fall when it starts to get cold I stick in the fork and pry up the leaves releasing a puff of steam from the heated interior. Nature is doing its work for me.

Through the winter months I continue feeding the first heap, the one with all the leaves raked from the yard. I clear a space in the snow on top of the leaf pile, dig down a foot or so with a fork and dump in the vegetable trimmings that have accumulated. I sprinkle lime on the garbage and cover it up with leaves again. Adding vegetable and plant matter goes on through all four seasons, so the heap is a continuing chore, always silently active and somewhere in the back of my mind. My attraction to it, my compulsion to do it, is as constant as the heaps are themselves. If I have some extra mulch hay I spread it over the two older heaps in the fall to protect them from the weather, but it isn't really necessary, it's just my fretting over them. After that they sit untouched all winter, ready in spring when everything starts up again.

11. Make-a-bleeve

Blocks was a made up activity, "make-a-bleeve" as some of the kids on Forest Hill Drive called "make believe" stories. The tricks that let us escape most completely can't be real. Real is what you fall back into when the spell ends. For the person I was at nine or ten, the hours spent building a miniature fantasy place with Philip were more completely absorbing and transporting--

But I've already described this and it's only an indulgent luxury to give myself the pleasure of glimpsing those humbly realized fantasies again. Although this time I can't help recalling both the intoxicated feeling they filled me with and also its inversion--the emptiness I felt when they were gone. This was how I discovered nullity. Some years later, a fledgling adult, I read a description of this feeling of nullity--the absence of any desire or savor for life. The writer was Camus. I'd read nihilism referred to before but this was the first I'd seen it described with such respect, the first I'd learned that this sensation of nothingness had a name, its own name, even had a genealogy and a dubiously distinguished history in philosophy. My sense of it had been a child's version, an immature hint of the most extreme fact of the problem of life. Sitting at the bar of the Elite one Wednesday night, Roland mentioned Camus' book *The Rebel.* Not a revolutionary, a rebel, he emphasized. A rebel is one who insists on freedom, for himself, for everyone. But he's not a revolutionary who wants to blow up the world to create a new one. The rebel's desire for freedom is individual, and his freedom has limits. Total freedom of the revolutionary kind leads to nihilism, the end of meaning, the triumph of the absurd. Camus was constructing his philosophy on what he saw as the absurd condition human beings found themselves in, alone in a silent universe, each of us continually pushing our rock up the hill like Sisyphus. "The absurd is born of the confrontation between human need and the unreasonable silence of the world," Roland quoted. How could you lead a life with any meaningful values, any values or meaning at all, in the face of this ultimate silence? Yet expressing it could give one a strange sense of assurance. It put you inside the fate of our species with a new awareness, even if it could also make you feel the immovable limits surrounding us. It was strangely empowering to know that serious people had grappled with the problem of what to do, how to live, if you could no longer play blocks.

In his usual emphatic manner, the manner of the philosopher, Roland re-defined words and ideas to make them his own: Camus' rebel, or perhaps it would be more accurate to say Roland's portrait of the rebel, is not the same as a revolutionary, a distinction he did not want me to miss. The latter is political and historical, his goal is to overturn everything, to create a new order. One who rebels is not allegiant to dogma but is insisting instead on the freedom of the individual, a freedom we must insist on otherwise the totalitarians and all their lesser incarnations descending from evil down to the merely destructive will win. Yet with freedom comes responsibility. So the circle closes. Camus' philosophy is not destructive but creative: "A nihilist is not one who believes in nothing, but one who does not believe in what exists." In other words, he rejects reality. Reality is the street you live on, the one where the maple trees grow. Camus had confronted the absurdity (to use his word) of man's lonely existence in a silent universe, a place recently evacuated by the death of God, and illuminated by the discovery that such make-a-bleeve was no longer possible, and we had no choice but to grow into our knowledge that we were absurd, our situation was absurd, yet we might allow the empty cup of meaninglessness to pass by barely touching our lips. "That was the kiss of recognition," Roland said in his own voice. "The recognition that there are limits and to go beyond them..." Roland paused, his eyes drilled me with their intensity, and he finished the thought: "...then we're fucked." Another pause and he exploded with his Henry the Eighth guffaw. The opening had been too tempting; he couldn't pass up mocking himself, mocking all seriousness in the face of the absurd. It was the fierce intensity in his eyes I was meant to believe.

What probably attracted Roland to Camus most strongly was his ambivalence toward and ultimately his rejection of all ideologies, including Communism. It was the dominant belief system, or religion, in the world left after the death of God. Camus had been born in 1913, and lived during World War Two in occupied France, editing a newspaper for the resistance, and was one of the writers defining the moral landscape after the war; he was awarded a Nobel Prize for his writing in 1960. Three years later he was killed in a car accident. An absurd death because it was meaningless, by Camus' own definition. His life was unique yet emblematic of his times. Everything in our century returned to the failure of humanity to resist the temptation to power, to rule itself justly and decently. That's the kind of thing one can say out here in space where our species, viewed from such a distance, can be thought of as a single organism with a sole mind capable of managing its own behavior. Which we can't. Not without inflicting damage on everyone else. Damage the rebel in all of us must object to. Now the danger is bringing the earth down on our heads, or its waters up our noses, washing ourselves and our civilization into the sea.

Before I met Roland, Camus was another name among the many I was hastily learning while trying to create a world in my mind that could include both make-a-bleeve and reality. After Roland's introduction, Camus became a part of who Roland was as well as taking up a place of his own. The death of deities is no longer shocking news, nor is the silent universe, and sometimes we are even able to laugh at our absurdity. Camus' thoughts have become part of the air we breathe. He gave us a rationale and a strategy for believing in an earthly life, but perhaps he has since been thrown on that growing compost heap of dead white men himself. After all, even if he was raised by a single mother in a poor Algerian neighborhood (his father had been a farm worker killed in World War One), he was French, a Pied-Noir, part of the colonial class, criticized in his time for not denouncing the French during the Algerian war for independence. To be part of the ruling class, even at the bottom, is to be, in another paraphrase of Camus, at least a little bit guilty.

12. Digression

Intruding itself on my view from the pyramid more than once was something I definitely did not want to see. To try to describe it would send me off on a major digression, one which might blot out everything else I've been able to see from this vantage. Does it matter? Well, nothing matters, for that matter, and if I don't want to do it, why would I? To get it off my conscience? It hurts my pride or vanity more than my conscience. So it's just selfish. I want to spare myself, spare that particular molted self the embarrassment, even if I have to admit to the same self that I'm close to reaching the point where nothing is embarrassing any more. (Not true at all, not even tru-ish.)

How bad was what I did, really? I was young, a little crazy, full of myself, and self-righteous at times. In other words, a certain type, the kind who overplayed his hand and screwed up a significant part of his life. I also let down a woman who had believed in me, not by being unfaithful but by breaking faith with her trust--the words required to confess it would only betray her again if I let them appear here in front of me. I grew out of that person, molted him, perhaps only in the mercy of time; time forgives everything, then with its last stroke it kills us, like a cat who's been teasing its prey. Yet it doesn't forgive anything at all, that's just sentimentality; only someone else can forgive.

Those years were lived in willful confusion as I underwent a peculiar evolution, blundering through it, surviving on momentum because I had a pen in my hand. I've re-read that sentence several times and still don't know what it means or where it came from. Maybe I've been too affected by reading philosophy and it means the opposite of what it says. To say the opposite would be meaningless. What it means contradicts what I said above, that nothing matters. With a pen in hand, everything matters, and that is the point of sitting out here on my perch in space. I'm trying to put enough space between me and my terrestrial home so that everything *doesn't* matter so much but I'm aware that this is merely another of those sleight-of-hand tricks that often fails to sustain its illusion after the flash.

I'm going to quit trying to understand these opaque comments and say something undeniable: Whomever each of us thinks we are, we are physical beings first and if we were truly able to understand the nature of the creature we inhabit we would probably find the person an unlovable disappointment whose company we wished to avoid. Unless we were blessed by the grace of a Mister Rogers, who is forgiveness personified. Nevertheless, it's too late, it's far too late, there is nothing to be done, not now, and some things can only be lived with in silence because there is no one to tell them to. Not even Anonymous wants to hear them.

13. Shhhh!

Almost sixty years ago Jerry and Eli stood on a golf course under a starry sky which was spinning in slow motion. Five thousand or so years before that Khufu's masons and workers, watched by a line of crows perched on a wall, were building a tomb to protect the pharaoh on his journey through eternity. Sixty-six million years before Khufu was put to rest in his pyramid, an asteroid collided with earth and wiped out the dinosaurs, and nearly fourteen billion years before that unwitnessed collision a speck, a less than visible dot of matter or energy, a mere physical instant which we have no name for began expanding into an unimaginably vast, unmapable, still-expanding space of time we call the universe.

The only one of these events that occupied the shape of an experience in Eli's or Jerry's mind was the most recent, the night on the golf course, which Eli remembered with the same baffled bemusement he felt at the time. A quite different kind of bemusement from what he felt at a temporally related moment in space/time when he was sweeping up the vestibule in The Lively Arts coffee house. Anchel and Sid had just come in, climbed the short flight of stairs to the main room and sat down at the host's table by the cash register. They were talking business and hadn't noticed Eli when they came in, or so Eli assumed, since Sid was in the middle of describing some deal he was involved in and Anchel hadn't said hello, although Anchel often moved in and out of occupied rooms without speaking. Sid was short and rounded, like Humpty-Dumpty, and always smoked a cigar. He talked all the time and didn't waste time being polite. He had a second hand furniture store and could get anything for you, but "nothing illegal, I'm not going to jail for you." It sounded as if Anchel had sent someone to Sid for something and the guy expected a better price than Sid could offer. Sid wanted Anchel to agree the price he named and was holding out for was fair. "You know me, Anchel. I'm here to do business. I never squeeze a deal too hard."

For Eli it became a slightly awkward moment. He was nearly done sweeping and still had to go up and clean the bathrooms but he didn't want to interrupt a private conversation, especially if they hadn't noticed he was there and had already said things they might not want him to hear. He crossed over to the other side of the vestibule, which went the width of the small building, thinking maybe they'd see him but he wouldn't be interrupting. He could smell Sid's cigar and remembered the time when the woman who lived upstairs shouted down to Anchel to close the window because the cooking smell was killing her. It was the corned beef he made once a week or so. The pungent fumes were being sucked out by the kitchen fan. "I'll call the landlord and complain!" she shouted. Sid was the landlord and Eli almost laughed out loud remembering Anchel murmuring, "Yeah, the landlord, call the landlord." Complaints were the background noise of Sid's life which he heard with the deafness of God.

"Her name is Sophie," Sid was saying, but he wasn't talking about his tenant. He had moved on to something else. "Built like a draft horse but what the hell, a hole's a hole. They're all the same, Anchel, some bigger, some smaller that's all."

"Shhh!" Anchel shushed him. "The kid--he doesn't know."

It's funny how I can hear this more clearly now than when I was standing there holding a broom, sweeping the tiles in the vestibule for the second time so I wouldn't interrupt Anchel and Sid. Now it was Anchel who was protecting me. But maybe it was something besides my innocence that Anchel wanted to protect. Maybe it was his own romantic ideal of the son he never had.

14. City-States of America

I can barely remember the sentimental map of the forty-eight continental states I carried around in my head during those early years, but it comes right back when I think of those forty-eight disuniting, pulling apart, as if they were taffy. My adult view of the map would like to do just that--pull it apart, carve it up into pieces. The era of the white man is over. John Betts with his black homburg and topcoat, his wave and confident greeting, the revered Past Exalted Ruler of the Elks Club striding through the center of the universe on his way to work is long in his grave and while remnants of his ilk remain, many in roles as masters of our universe, the future is not theirs. It's not mine either. I can now see what a mistake it was to become infatuated with the notion of the "original" forty-eight states as a formal, idealized construction. It was the mental act of a boy, seduced by size and strength, as if the

"Forty-eight" was a sports team I was rooting for. Size is the problem, the way it enables the will to power, pumps us all up. And if we're honest we have to admit it was necessary for us to push people out of our way right from the start--how else could we settle here when they were here first! Embarrassment over this brute bloody fact has been part of our birthright, as feebly as it has stirred our national conscience. That's how the transcontinental nation got its first push on the way to becoming the American empire, a malignant, metastatic state which has led us to commit crimes on a global scale, from Sand Creek to My Lai. We have done this almost helplessly, as if size itself tempts the nation to assume a predatory personality so it becomes inevitable that we get Andy Jackson and Teddy Roosevelt as national heroes and presidents. And in a later age, seduced by their own imaginations as surely as I was at ten, an actor president and a celebrity one. Neither ever served in a role where he risked getting one of our nation's real victim's blood on his own hands. Like many of the rest of us, they have been or are able to live their long lives believing in their own innocence.

Yet I am one of them too. What else could I be if I'm not an American, a son of these fifty states? At my age I'm stuck here, living out my final years in futile frustration with what the fifty have done and what we've become, while also enjoying my portion of the empire's wealth. This is the hypocrisy we all support, the stain that seeps into everyone sharing our address and our history.

As a last gesture and final goodbye to my childhood fantasy, I'm going to re-imagine the map in an alternative arrangement to the sea-to-shining-sea forty-eight. To achieve this, I'm using the model of the "city-state" which Mark Peterson describes in his history, *The City-State of Boston*. (Professor Peterson is a serious historian and is not to be blamed for inspiring my fantasies.) This new form will divide and collect the states into smaller polities, as if they were drawn together magnetically around several concentrated urban areas. My choices for making this new geographic arrangement are based on geography and culture, including history. An element of arbitrariness is inevitable. Obviously not all these polities are the same size, and there will be other inequalities too. There always are. The goal is that by having smaller polities the affinities holding the "city-states" together as political units will be stronger and feel more natural to the citizens. As sovereign political entities I imagine them to be free to associate themselves with other polities as is useful or convenient, including changing the boundaries demarcating this new geography.

The city-state of Boston will comprise Massachusetts, Connecticut, Rhode Island, Maine, New Hampshire and Vermont. If green and crusty Vermont objects, let them go it alone, or consider joining themselves to Quebec or some part of Canada. Like everything else suggested here, it's up to them.

The city-state of New York will comprise New York and New Jersey.

The city-state of Philadelphia will comprise Pennsylvania and Delaware.

The city-state of Virginia will comprise Virginia, West Virginia, Maryland, and the District of Columbia.

The city-state of Atlanta will comprise Georgia, North Carolina, South Carolina, Alabama and Mississippi.

The city-state of Orlando will comprise Florida (Miami will soon be flooded and uninhabitable) and Puerto Rico.

The city-state of Nashville will comprise Tennessee and Kentucky.

The city-state of Baton Rouge will comprise Louisiana and Arkansas. (New Orleans will also soon be mostly uninhabitable.)

The city-state of Kansas City (on both sides of the river) will comprise Missouri, Kansas, Nebraska, and Oklahoma.

The city-state of Chicago will comprise Illinois, Indiana, Ohio, and Michigan.

The city-state of Minneapolis-St. Paul will comprise Minnesota, Iowa, and Wisconsin.

The city-state of Missoula will comprise North Dakota, South Dakota, Montana, Wyoming, and Idaho.

The city-state of Salt Lake City will comprise Utah (this is the land of Deseret they always wanted).

The city-state of Denver will comprise Colorado, Arizona, and New Mexico.

The city-state of Sacramento will comprise California, Oregon, Washington, and Nevada.

The city-state of Austin will comprise Texas.

That's the forty-eight continental states. Anchorage can comprise the city-state of Alaska, and the city-state of Honolulu comprises the islands of Hawaii.

As I said, my goal is to think in a way that does not revolve around a transcontinental nation. These monstrously sized polities are the problem. They have so much weight that they can't help throwing it around. To be consistent and do the job right we should pull apart Russia (already less monstrous than the Soviet Union) and China too. Also India. But who is this omnipotent *we*? It's the reader's conscience multiplied by a large unknown factor. If a time comes when due to advances in physics and biology, it is possible for us to reify this multiple quantity of the human conscience into an actual organic form, then wonderful, we might enter a new moral universe. If not, we still have an ideal to aim at.

Our federal system makes it quite natural to carve the states up into more manageable units; smaller political entities would be less prone to global mischief, more inclined to live peaceably with their neighbors, and better able to focus on the abiding purpose of government anywhere--to treat its citizens decently. (I take this commandment from two twentieth century authorities, George Orwell and Albert Camus).

Of course, as I acknowledged, nothing like this is going to happen. My scheme is impossible, and can only exist as a symbol standing for what the polities represent--a mental figment like the original forty-eight state flag that flew in my child's mind. This one is the flag of the City-States of America, not joined in political union but related by sharing the ideal named by both Orwell and Camus--to treat their citizens decently.

15. *First day of spring*

In the Philippines a dead whale washed up on a beach. Local authorities performed an autopsy and discovered eighty-eight pounds of plastic in the whale's stomach. The plastic had been there so long it was compacted into a solid mass which the whale couldn't expel. The ingested plastic takes up so much space in the whale's stomach, said the scientist, the whale thinks it is full even when starving, and they often become weak from lack of nourishment, sicken, and die.

Well, we all wash up on some beach eventually, so what's the big deal? At least a few of us will end in some horrific way. Yet what anyone is doing here--on a beach, in the ocean, up on land--is a question whose meaning is annihilated by an end like the whale's. Not having died yet, but having seen my father quietly expire, and having had a foretaste of oxygen deprivation due to angina, with its whiff of basal fear, my strong guess is that at the final moment we do not enjoy a conclusive view of our whole collage. That is movie stuff. If you're starving or gasping for air what goes through your mind is not picturesque. So perhaps the best time to see the whole thing is right now, whenever now might be. Even if all we get is a glimpse and it can never be seen whole. I've spent way too much time trying to figure things out that can't be figured out, but even now, knowing the futile temptation, I can't help it. I think a stomach full of compacted plastic would cure me fast.

If a collage is flat, all centers, without an end or a beginning, a perfect democracy of scenes, none with a privileged view, then it doesn't much matter where you look first, or last. (Or, as Ring Lardner might have said, you can skip them all at once.) But the written, the spoken word, also has an arrow pulling it along, and whether you like it or not the arrow has its own momentum, expressed in motion, which is the subject of the quotation I was going to use as an epigraph. In his *Essays,* Montaigne wrote: "All things are in constant motion, the earth, the rocks of the Caucasus, the pyramids of Egypt, both with common motion and with their own. Stability itself is nothing but a more languid motion." I put this quotation here instead of at the top because I've noticed from my own experience that starting with an epigraph a reader can't really know why it's relevant to the pages that follow. It's just an ornament hung on the book to give it a tone of significance. I was going to drop it because of its pretentiousness but I'm leaving it in because of the coincidental mention of pyramids and its reference to motion which is a form of time--time made manifest--and the underlying subject of my collage, if it has one.

16. A rotting turnip

No matter what we do, even without the addition of more violent rain storms and fatter, hungrier pests, the garden comes to an end every fall. Nothing is more certain. There's always a scent of loss in the air along with relief that the heavier work will soon be done. For Kate the end of the gardening season means planting next year's garlic; she does this in October. For me it's moving the compost heap. As I said in my description of rotating three heaps, I don't turn them during the summer months; there are too many other things to do. But if you don't turn them some time, the bacteria won't get the oxygen they need to break down all the organic matter. So now is the time.

While shoveling compost, I can look up in the sky just in case there might be a pyramid gliding by. It hasn't happened yet but as David Hume said about "matters of fact," just because something has never happened doesn't mean it never will. As a practical matter we assume it won't because we know pyramids don't fly, but we can't prove it. We can only say it's never happened that we know of. Hume's description of the limits of inductive reasoning is a neat demonstration of the illogic of logic, of the way it can defy common sense by claiming a superior truth. He also provides a speculation that makes me think again of the strange tug the compost heap exerts on the human being who is tending it. "All the operations of nature," Hume says, "whether the rotting of a turnip, the generation of an animal or the structure of human thought probably bear some remote analogy to each other." This speculation was all intuition, of course, something else he couldn't prove, but to find a rotting vegetable, a newborn creature, and the structure of thought linked in this way was the closest I would feel, in my own evolved state, to a sacred connection with the rest of life on earth.

17. Many worldz two

Where I have ended up, after seventy-six years on earth (some time has passed since I began writing these scenes and notes) is in physics. In this location, I'm like a person who has gone for a walk in the woods, became distracted by the growing things around him and also by the randomly summoned thoughts in his head, and is now lost. Not utterly lost, because I know which direction I have to walk to get back home, though I don't know how long it will take, and if I get distracted on the way back, as I usually do, I may get lost again and have to backtrack a second time. So this could be a long walk. The sense that I know roughly but not exactly where I am is a good metaphor for my life. As I've grown older I could say I've learned to recognize my personal routes through the woods better so I don't get so completely disoriented and off track as I did years ago but I'm still stumbling around at times with the frustrating feeling I've been here before yet I still don't know just where this is.

The one thing reading Lee Smolin's account of his attempt to complete Einstein's revolution has done is to confirm my intuition that there is one universe and it is real. So, he says, is time, as I quoted him saying earlier: "Indeed, our experience of time's passage is the one thing we directly perceive about the world which is truly fundamental. All the rest, including the impression that there are unchanging laws, is approximate and emergent."

I've learned a lot reading Dr. Smolin. I'm not going to tell you what else I learned and risk revealing I misunderstood him, but I can say something about the experience. A couple times reading *Einstein's Unfinished Revolution* (subtitled, *The Search for What Lies Beyond the Quantum*) I had that suddenly elevating sensation of understanding something new about the cosmos. The glimpse had a clarity and force that made me feel I was close to the reality of existence. Among the things I owe Dr. Smolin is the confirmation that I'm right to believe in reality. That's not a wisecrack. The twentieth century, where I lived until I was a middle aged man, has discredited reality. Some people would probably say this is due to our having had to absorb too much reality in so many shocking and unsettling forms during that turbulent span. The atomic bomb can stand for all the shattering realities that the century, in response to our prying and probing using science, has disclosed to us evermore. Well, maybe not evermore. No one knows where this is going. Signs for our survival are mixed, and trending downward.

As for me, I'll be leaving soon. I'm not going to learn everything I'd like to know but Dr. Smolin hints the same thing in his book, and Einstein himself was unable to finish the revolution in human knowledge he began in 1905. Even the most luminous minds have their limits. That can be a sad fact to contemplate, but it can also be comforting in an odd way to think that my frustration in living with gaps in comprehension of the very weird and hard to believe universe is not entirely unlike the gaps Einstein and Dr. Smolin themselves experience(d) when they lay their heads down on the pillow. Yes, their ignorance is far richer than mine, and when they visit what they know they can stay for hours, days, months, while I am gently guided out the door when it feels as if I just arrived. But the core experience, that one will never know enough to be satisfied, is the same. To be human is to live with unsolved puzzles in one's head.

18. A hungering will

There's always more to say but it's time to lapse into silence again and enjoy the view from my perch on the oh-so-weightless stone tomb floating out here in space. I know I won't be able to enjoy the trip forever; at some point my eyes will close and everything will vanish. The pyramid will hover for a moment, a fugitive from gravity, quite unaware that before it became a tourist attraction and an astronomical anomaly it was the home of a dead pharaoh. Poor clunky tetrahedron, adrift in space. What if it was the only trace left of humans' time on earth? What kind of story could a wondering traveler make out of it? My guess is that it might be a very strange one but nowhere near the truth, which as Steven Weinberg said about the world the physicists keep discovering for us often turns out to be weirder than anyone could imagine.

Right here we should have a song, to take us out. Call it the *Pyramid in Space* theme song. The notion we want the composer and lyricist to be inspired by, or simply to start from if they need a suggestion, is that looking for cosmic significance out there somewhere is looking in the wrong place (good line for the title). We are the source of meaning. Without us the universe merely is, an evolving physical fact, a cosmic fireworks display. Passed through our blood, sweat, toil, and tears, expressed by our voice, recorded by our memory, what we call meaning may emerge. Or it may not. A hungering will, to use Roland's phrase, is also necessary.

We will give our composing team some time (and space) to come up with a song (maybe one the whales will enjoy too), and in the meantime mull what this is all about. It sounds a little silly of course. So many people have tramped over this ground that nothing can grow here and the sound we hear in the background is a stadium full of teenagers laughing at the absurdities they are freshly discovering about bipedal life. Well, yes. It is funny when you think the most human question is such a creaking cliche. You can't even say it with a straight face. Then the next revelation, or perhaps it's only an insight (still waiting for the first draft of *Looking in the Wrong Place* to be handed in*)*, that all our stories, all our rituals and songs we've made up to try to say it--

Now I've forgotten what I was going to say that felt so important. No matter. It will come to me later.

END